THE MAGICIAN'S TWIN

The Magician's Twin

C. S. Lewis on Science, Scientism, and Society

John G. West, Editor

Seattle · Discovery Institute Press · 2012

Description

Beloved for his Narnian tales and his books of Christian apologetics, bestselling author C.S. Lewis also was a prophetic critic of the growing power of scientism, the misguided effort to apply science to areas outside its proper bounds. In this wide-ranging book of essays, contemporary writers probe Lewis's warnings about the dehumanizing impact of scientism on ethics, politics, faith, reason, and science itself. Issues explored include Lewis's views on bioethics, eugenics, evolution, intelligent design, and what he called "scientocracy."

Copyright Notice

Copyright © 2012 by Discovery Institute. All Rights Reserved.

Publisher's Note

This book is part of a series published by the Center for Science & Culture at Discovery Institute in Seattle. Previous books include *The Deniable Darwin* by David Berlinski, *In the Beginning and Other Essays on Intelligent Design* by Granville Sewell, *God and Evolution: Protestants, Catholics, and Jews Explore Darwin's Challenge to Faith*, edited by Jay Richards, and *Darwin's Conservatives: The Misguided Quest* by John G. West.

Library Cataloging Data

The Magician's Twin: C. S. Lewis on Science, Scientism, and Society

Edited by John G. West. Foreword by Phillip E. Johnson.

Contributions by M. D. Aeschliman, Edward J. Larson, Jake Akins, C. John Collins, Jay Richards, Victor Reppert, James Herrick, Michael Miller, and Cameron Wybrow.

348 pages, 6 x 9 x 0.94 inches & 1.5 lb, 229 x 152 x24 mm. & 0.67 kg

Library of Congress Control Number: 2012942135

BISAC: SCI075000 SCIENCE / Philosophy & Social Aspects

BISAC: REL106000 RELIGION / Religion & Science

BISAC: BIO007000 BIOGRAPHY & AUTOBIOGRAPHY / Literary

ISBN-13: 978-1-936599-05-9 (paperback)

Publisher Information

Discovery Institute Press, 208 Columbia Street, Seattle, WA 98104

Internet: http://www. discovery.org/

Published in the United States of America on acid-free paper.

First Edition, First Printing. September 2012.

For Sonja, Katherine, and Garrett.

EXPLORE MORE

Access videos, articles, and additional free resources about C. S. Lewis, including *The Magician's Twin* documentary.

Contents

Foreword 9
   Phillip E. Johnson

Introduction 11
   John G. West

Science & Scientism

1 **The Magician's Twin** 19
   John G. West

2 **C. S. Lewis on Mere Science** 47
   M. D. Aeschliman

3 **C. S. Lewis on Science as a Threat to Freedom** 53
   Edward S. Larson

4 **C. S. Lewis, Science, and the Medieval Mind** 59
   Jake Akins

5 **A Peculiar Clarity** 69
   C. John Collins

Origins

6 **Darwin in the Dock** 109
   John G. West

7 **C. S. Lewis and Intelligent Design** 153
   John G. West

Reason

8 **Mastering the Vernacular** 179
   Jay W. Richards

9 **C. S. Lewis's Dangerous Idea Revisited** 199
   Victor Reppert

SOCIETY

10 C. S. LEWIS AND THE ADVENT OF THE POSTHUMAN 235
    James A. Herrick

11 THE EDUCATION OF MARK STUDDOCK 265
    Cameron Wybrow

12 C. S. LEWIS, SCIENTISM, AND THE BATTLE OF THE BOOKS . . . 293
    M. D. Aeschliman

13 C. S. LEWIS, SCIENTISM, AND THE MORAL IMAGINATION . . . 309
    Michael Matheson Miller

BIOGRAPHIES OF CONTRIBUTORS 339

INDEX 343

Foreword

Phillip E. Johnson

C. S. Lewis's many admirers will be eager to read this collection of articles, collected and edited by the Discovery Institute's John West, concerning Lewis's views of science, which he respected, and scientism, against which he warned. The book contains a timely and well-reasoned chapter about Lewis and intelligent design, which seems to have replaced creationism as the alternative most feared and reviled by Darwinists. Another chapter describes the subtle interconnection between *That Hideous Strength* (my favorite Lewis novel) and his much admired philosophical work, *The Abolition of Man*.

As West notes in chapter 1, C. S. Lewis remarked that "[t]he serious magical endeavour and the serious scientific endeavour are twins"—an image that gives this book its title. Lewis meant that modern science and magic have a common starting point in history, arising from efforts to understand and manipulate nature, and they have retained important and perhaps unexpected similarities down to the present. His point makes me think of what scientific studies of identical twins, separated at birth and raised apart, have shown. Such studies consistently demonstrate that, aside from physical resemblance, when the twins first meet each other decades later, they display striking similarities in matters so unexpected and detailed that they seem eerie. It is as if the studies were aimed at proving that, despite all we have learned about stars since 1600, astrology may nonetheless still have an impressive power of prediction.

Of course, the twin studies support genetics, not astrology, but what they teach us about identical twins raised apart makes it unsurprising that the scientific culture of the nineteenth and early twentieth century produced three great wizards—Charles Darwin, Karl Marx, and Sigmund Freud—whose concepts were so spellbinding that they set the intellectual agenda for the entire twentieth century. In many ways, they still hold us in their grasp.

I would add to the list of scientific magicians the "DNA is everything" biologists, including the brilliant popularizer Richard Dawkins and the physicalist neuroscientists who assure us that our thoughts and decisions (including the conclusions of neuroscientists?) are no more than the effects of electro-chemical events in the brain. These have sought to make science indistinguishable from scientism, and thus have inadvertently alerted us to the continuing importance of C. S. Lewis's exposure of the irrationality of scientism.

Overall, this collection charms the reader, not because Lewis has necessarily said the final word on every subject covered, but because his perceptive words illuminate every subject and inspire discussion in which participants can employ their own intellects to move ever closer to the truth.

Introduction

John G. West

Narnia. Screwtape. *Mere Christianity*. With more than 200 million copies of his books reportedly sold, C. S. Lewis is known and beloved by readers around the globe for his children's stories, his works of theology, and his winsome (and witty) defenses of orthodox Christianity.[1]

One thing Lewis is not particularly well-known for is his views on science.

Yet he ultimately wrote nine books, nearly 30 essays, and several poems that explored science and its cultural impact, including *The Discarded Image*, his last book, which critically examined the nature of scientific revolutions, especially the Darwinian revolution in biology.[2] Lewis's personal library, meanwhile, contained more than three dozen books and pamphlets on scientific subjects, many of them dealing with the topic of evolution. Several of these books were marked up with underlining and annotations, including Lewis's copy of Charles Darwin's *Autobiography*.[3]

Throughout his life, Lewis displayed a healthy skepticism of claims made in the name of science. He expressed this skepticism even before he was a Christian. For example, while still an unbelieving undergraduate in 1922, he recorded in his diary a discussion with friends where they expressed their doubts about Freud.[4] In 1925, he wrote his father about his gratitude toward philosophy for showing him "that the scientist and the materialist have not the last word."[5] The next year he published his narrative poem *Dymer*, which offered a nightmarish vision of a totalitarian state that served "scientific food" and "[c]hose for eugenic reasons who should mate."[6]

In 1932, just a few months after becoming a Christian, Lewis wrote to his brother about the efforts of the Rationalist Press Association to publish cheap editions of scientific works they thought debunked religion. Lewis said their efforts reminded him of the remark of another writer "that a priest is a man who disseminates little lies in defence of a great truth, and a scientist is a man who disseminates little truths in defence of a great lie."[7]

During the 1940s and 1950s, Lewis became more vocal about the looming dangers of what he called "scientocracy," the effort to hand over the reins of cultural and political power to an elite group of experts claiming to speak in the name of science.[8] Lewis regarded this proposal as fundamentally subversive of a free society, and he worried about the creation of a new oligarchy that would "increasingly rely on the advice of scientists till in the end the politicians proper become merely the scientists' puppets."[9]

Lewis took pains to emphasize that he was not "anti-science."[10] But he unequivocally opposed *scientism*, the wrong-headed belief that modern science supplies the only reliable method of knowledge about the world, and its corollary that scientists have the right to dictate a society's morals, religious beliefs, and even government policies merely because of their scientific expertise.

Because Lewis died nearly five decades ago, we might be tempted to think that he inhabited a vastly different world than we do when it comes to the relationship between science and culture. But in key respects, Lewis's world was very much like our own. Then, as now, certain prominent intellectuals claimed that science provides a view of the universe that refutes the traditional religious view. Then, as now, certain pundits claimed that you were "anti-science" merely for being skeptical of certain claims made in the name of science. And then, as now, some spokespersons for the scientific establishment insisted that public policy should be guided—even dictated—by an elite class of "scientific" experts.

As the essays in this book show, Lewis has important things to tell us about the limits of science, the need for dissent in science, and the dangers of trying to govern in the name of science. Along the way, Lewis offers penetrating insights into many hot-button issues of our time, including evolution, intelligent design, bioengineering, moral relativism, and even the role of government.

Consider this book an invitation to think more deeply about the growing power of science in the public square by drawing on the timeless wisdom of C. S. Lewis. After you have read the book, I encourage you to avail yourself of the additional articles, companion videos, and other resources at the website www.cslewisweb.com, including a new documentary film about Lewis and scientism inspired by this book.

Appreciations

The Magician's Twin could not have been completed without the help of a number of key people and groups. I especially want to thank the Maclellan Foundation for providing the grant that made both the book and documentary possible; Charlie Phillips at the Maclellan Foundation for encouraging me to pursue the project; Jake Akins for his assistance with research into unpublished Lewis material held at the Wade Center; the staff of the Wade Center at Wheaton College for graciously facilitating some of the research for this volume; Cameron Wybrow for his meticulous proofreading; my assistant Jens Jorgenson; and the writer Michael W. Perry for his expert skills in typesetting the book.

The essays published here are new except for two: chapter 2, "C. S. Lewis on Mere Science," which originally appeared in the journal *First Things* (www.firstthings.com), October 1998, No. 8 and is reprinted here by permission; and chapter 3, "C. S. Lewis on Science as a Threat to Freedom," which was commissioned for a conference on C. S. Lewis and Public Life held by Discovery Institute in 1995 and appears here in book form for the first time.

The brief quotations from C. S. Lewis's published works cited in this book are copyright by the C. S. Lewis Company and are used under the terms of the fair use provisions of the copyright law. Quotations of unpublished Lewis material in chapter 6 are also copyright by the C. S. Lewis Company, and are used here by permission of the Lewis Co. and the Wade Center at Wheaton College.

Endnotes

1. "Wheaton College to Screen C. S. Lewis Documentary," *The Daily Herald*, October 20, 2001, accessed June 5, 2012, http://www.highbeam.com/doc/1G1-79384514.html.
2. Books by Lewis that have a major focus on science and its relationship to culture include: *The Pilgrim's Regress* (1933); *Out of the Silent Planet* (1938); *Perelandra* (1943); *That Hideous Strength* (1945); *The Problem of Pain* (1940); *The Abolition of Man* (1944); *Miracles: A Preliminary Study* (1947); *The Magician's Nephew* (1955); *The Discarded Image* (1964). Essays by Lewis with a major focus on science include: "De Futilitate" (1940s), "Funeral of a Great Myth" (probably 1940s); "Bulverism" (original version published in 1941; expanded version in 1944); "Miracles" (1942); "Dogma and the Universe" (1943); "Horrid Red Things" (1944); "Religion and Science" (1945); "Is Theology Poetry?" (1945); "The Laws of Nature" (1945); "Christian Apologetics" (1945); "Two Lectures" (1945); "Man or Rabbit?" (circa 1946); "Religion without Dogma?" (1946); "A Reply to Professor Haldane" (circa 1946); "Modern Man and His Categories of Thought" (1946); "Vivisection" (1947); "On Living in an Atomic Age" (1948); "The Humanitarian Theory of Punishment" (1949); "The Empty Universe" (1952); "The World's Last Night" (1952); "On Punishment: A Reply to Criticism" (1954); "On Obstinacy in Belief" (1955); "De Descriptione Temporum" (1955); "On Science Fiction" (1955); "Religion and Rocketry" (1958); "Behind the Scenes" (1956); "Is Progress Possible? Willing Slaves of the Welfare State" (1958); "The Seeing Eye" (1963). Poems broaching scientific themes include "The Adam Unparadised," "Evolutionary Hymn," "Prelude to Space," "Science Fiction Cradlesong," "An Expostulation," and "On the Atomic Bomb." Most of Lewis's essays are reprinted *God in the Dock: Essays on Theology and Ethics*, edited by Walter Hooper (Grand Rapids: Eerdmans, 1970); *Christian Reflections*, edited by Walter Hooper (Grand Rapids: Eerdmans, 1967); *Present Concerns*, edited by Walter Hooper (New York: Harcourt Brace Jovanovich, 1986); and *Selected Literary Essays* (Cambridge: Cambridge University Press, 1955). His poetry can be found in *Poems*, edited by Walter Hooper (San Diego: 1964) and *Narrative Poems*, edited by Walter Hooper (London: Fount Paperbacks, 1994).
3. These books are presently held at the Wade Center, Wheaton College. For a listing of the surviving books from Lewis's personal library, consult the description in "C. S. Lewis Library" (Wade Center, 2010), accessed May 18, 2012, http://www.wheaton.edu/wadecenter/Collections-and-Services/Collection%20Listings/~/media/Files/Centers-and-Institutes/Wade-Center/RR-Docs/Non-archive%20Listings/Lewis_Public_shelf.pdf.
4. "We talked a little of psychoanalysis, condemning Freud." Entry for May 26, 1922, in C. S. Lewis, *All My Road Before Me: The Diary of C. S. Lewis, 1922-1927*, edited by Walter Hooper (San Diego: Harcourt Brace Jovanovich, 1991), 41.
5. C. S. Lewis to his Father, Aug. 14, 1925 in *C. S. Lewis: Collected Letters*, edited by Walter Hooper (London: HarperCollins, 2000), vol. I, 649.
6. C. S. Lewis, "Dymer"(1926), *Narrative Poems*, 7, 20.
7. C. S. Lewis to Warren Lewis, April 8, 1932, *The Collected Letters of C. S. Lewis* (San Francisco: HarperSanFrancisco, 2004), vol., II, 75.
8. C. S. Lewis to Dan Tucker, Dec. 8, 1959, in *The Collected Letters of C. S. Lewis*, edited by Walter Hooper (San Francisco: HarperSanFrancisco, 2007), vol. III, 1104.

9. C. S. Lewis, "Is Progress Possible? Willing Slaves of the Welfare State," *God in the Dock*, 314.
10. C. S. Lewis, *The Abolition of Man* (New York: Macmillan, 1955), 86.

SCIENCE & SCIENTISM

It might be going too far to say that the modern scientific movement was tainted from its birth: but I think it would be true to say that it was born in an unhealthy neighbourhood and at an inauspicious hour.

—C. S. Lewis, *The Abolition of Man*

1

The Magician's Twin

John G. West

In his classic book *The Abolition of Man* (1944), C. S. Lewis wrote that "[t]he serious magical endeavour and the serious scientific endeavour are twins."[1]

At first reading, Lewis's observation might seem rather strange. After all, science is supposed to be the realm of the rational, the skeptical, and the objective.

Magic, on the other hand, is supposed to be the domain of the dogmatic, the credulous, and the superstitious. Think of a witch doctor holding sway over a tribe of cannibals deep in a South Sea jungle.

As strange as Lewis's observation might first appear, the comparison between science and magic runs throughout a number of his works. The sinister Uncle Andrew in Lewis's Narnian tale *The Magician's Nephew* is both a magician *and* a scientist; and the bureaucratic conspirators at the National Institute of Co-ordinated Experiments (N.I.C.E.) in Lewis's adult novel *That Hideous Strength* crave the powers of both science and the magician Merlin in their plot to reengineer society.[2]

For all of the obvious differences between science and magic, Lewis correctly understood that there are at least three important ways in which they are alarmingly similar. More than that, he recognized that these similarities pose a growing threat to the future of civilization as we know it.

1. Science as Religion

The first way science and magic are similar according to Lewis is their ability to function as an alternative religion. A magical view of reality can inspire wonder, mystery, and awe. It can speak to our yearning for something beyond the daily activities of ordinary life. Even in our technocratic age, the allure of magic in providing meaning to life can be seen in the continuing popularity of *Star Wars*, *The Lord of the Rings*, the Narnian chronicles, and the adventures of Harry Potter. While magical stories tantalize religious and irreligious people alike, for those without conventional religious attachments, they can provide a substitute spiritual reality.

Modern science can offer a similarly powerful alternative to traditional religion. In Lewis's lifetime, the promoter par excellence of this sort of science as religion was popular writer H. G. Wells. Wells and others fashioned Darwin's theory of evolution into a cosmic creation story Lewis variously called "The Scientific Outlook," "Evolutionism," "the myth of evolutionism," and even "Wellsianity."[3] While some contemporary evolutionists contend that people doubt Darwinian theory because it does not tell a good story,[4] Lewis begged to differ. In his view, cosmic evolutionism of the sort propounded by Wells was a dramatic narrative brimful of heroism, pathos, and tragedy.

In a bleak and uncaring universe, the hero (life) magically appears by chance on an insignificant planet against astronomical odds. "Everything seems to be against the infant hero of our drama," commented Lewis, "... just as everything seems against the youngest son or ill-used stepdaughter at the opening of a fairy tale." No matter, "life somehow wins through. With infinite suffering, against all but insuperable obstacles, it spreads, it breeds, it complicates itself, from the amoeba up to the plant, up to the reptile, up to the mammal."[5] In the words of H. G. Wells, "[a]ge by age through gulfs of time at which imagination reels, life has been growing from a mere stirring in the intertidal slime towards freedom, power and consciousness."[6]

Through the epic struggle of survival of the fittest, Man himself finally claws his way to the top of the animal kingdom. Eventually he finds Godhood within his grasp if only he will seize the moment. To quote Wells again: "Man is still only adolescent... we are hardly in the earliest dawn of human greatness... What man has done, the little triumphs of his present state, and all this history we have told, form but the prelude to the things that man has got to do."⁷

But then, after Man's moment of triumph, tragedy strikes. The sun gradually cools, and life on Earth is obliterated. In Wells's *The Time Machine*, the protagonist reports his vision of the dying Earth millions of years hence: "The darkness grew apace; a cold wind began to blow… From the edge of the sea came a ripple and whisper. Beyond these lifeless sounds the world was silent… All the sounds of man, the bleating of sheep, the cries of birds, the hum of insects, the stir that makes the background of our lives—all that was over."⁸

Lewis explained that he "grew up believing in this Myth and I have felt—I still feel—its almost perfect grandeur. Let no one say we are an unimaginative age: neither the Greeks nor the Norsemen ever invented a better story." Even now, Lewis added, "in certain moods, I could almost find it in my heart to wish that it was not mythical, but true."⁹

Lewis did not claim that modern science necessitated the kind of blind cosmic evolutionism promoted by H. G. Wells and company. Indeed, in his book *Miracles* he argued that the birth of modern science and its belief in the regularity of nature depended on the Judeo-Christian view of God as Creator: "Men became scientific because they expected Law in Nature, and they expected Law in Nature because they believed in a Legislator."¹⁰ Nevertheless, Lewis thought that biology after Darwin provided potent fuel for turning science into a secular religion.

One does not need to look very far to see science being used in the same way today. In 2012 thousands of atheists and agnostics converged on Washington, D.C. for what they called a "Reason Rally."¹¹ The rally had all the trappings of an evangelistic crusade, but instead of being preached at by a Billy Graham or a Billy Sunday, attendees got to hear Darwinian biologist Richard Dawkins and *Scientific American* columnist Michael Shermer. Former Oxford University professor Dawkins is known for claiming that "Darwin made it possible to be an intellectually fulfilled atheist," while Shermer once wrote an article "Science is My

Savior," which explained how science helped free him from "the stultifying dogma of a 2,000-year-old religion."[12]

The central role Darwinian evolution continues to play for the science-as-religion crowd is readily apparent in the countless "Darwin Day" celebrations held around the globe each year on Feb. 12, Charles Darwin's birthday. Darwin Day is promoted by a group calling itself the International Darwin Day Foundation. Managed by the American Humanist Association, the group's mission is "to encourage the celebration of Science and Humanity" because "[s]cience is our most reliable knowledge system."[13]

According to Amanda Chesworth, one of the co-founders of the Darwin Day movement, the purpose of Darwin Day is to "recognize and pay homage to the indomitable minds and hearts of the people who have helped build the secular cathedrals of verifiable knowledge." Chesworth's word choice is particularly astute: by doing science, scientists in her view are building "secular *cathedrals*."[14] The iconography of religion is unmistakable. In the words of one Darwin Day enthusiast who posted an approving comment on the official Darwin Day site: "To me, Charles Darwin is more of a God than the one armies had nailed to a cross."[15]

Perhaps the most tireless proponents of cosmic evolutionism today are the husband-and-wife team of Michael Dowd and Connie Barlow, who bill themselves as "America's Evolutionary Evangelists."[16] A former evangelical Christian turned Unitarian minister turned religious "naturalist," Dowd is author of *Thank God for Evolution!*, the subtitle of which is "How the Marriage of Science and Religion Will Transform Your Life and Our World."[17] Dowd calls his brand of cosmic evolutionism the "Great Story," which is defined on the Great Story website as "humanity's sacred narrative of an evolving Universe of emergent complexity and breathtaking creativity—a story that offers each of us the opportunity to find meaning and purpose in our lives and our time in history."[18] The Great Story comes along with its own rituals, parables, hymns, sacred sites, "evolutionary revival" meetings, Sunday School curricula, and even "cosmic rosaries," necklaces of sacred beads to teach children the fundamental doctrines of cosmic evolutionism.[19]

Dowd has attracted widespread support from Nobel laureates, atheistic evolutionists, and theistic evolutionists. For all of his outreach to the faith community, however, Dowd dismisses the reality of God just as much as atheist biologist Richard Dawkins. In an article written for *Skeptic* magazine, Dowd acknowledged his view that "God" is simply a myth: "God is not a person; God is a personification of one or more deeply significant dimensions of reality."[20] Just as people in the ancient world personified the oceans as the god Poseidon or the sun as the god Sol, contemporary people personify natural forces and call them "God."[21] Hence, Dowd's Great Story is ultimately a drama of the triumph of blind and undirected matter in a universe where a Creator does not actually exist. This becomes explicit in the description of the Great Story provided by philosopher Loyal Rue, cited approvingly on the Great Story website:

> In the course of epic events, matter was distilled out of radiant energy, segregated into galaxies, collapsed into stars, fused into atoms, swirled into planets, spliced into molecules, captured into cells, mutated into species, compromised into thought, and cajoled into cultures. **All of this (and much more)** is what **matter** has done as systems upon systems of organization have emerged over thirteen billion years of creative natural history.[22]

"All of this… is what *matter* has done," not God. Just like the narrative promoted by H. G. Wells and the scientific materialists at the beginning of the twentieth century, the cosmic evolutionism offered by Dowd and his followers in the twenty-first century is ultimately reducible to scientific materialism. The bottom line of their secular creation story is neatly encapsulated by Phillip Johnson: "In the beginning were the particles. And the particles somehow became complex living stuff. And the stuff imagined God, but then discovered evolution."[23]

Lewis would not have been surprised by current efforts to co-opt traditional religion in the name of science, or even to find a lapsed clergyman leading the charge. In Lewis's novel *That Hideous Strength*, the sometime clergyman Straik joins hand-in-hand with the avowed scientific materialists in the name of promoting a new this-wordly religion. As the impassioned Rev. Straik declares to Mark Studdock: "The Kingdom is going to arrive: in this world: in this country. The powers of science are an instrument. An irresistible instrument."[24]

2. Science as Credulity

THE SECOND WAY SCIENCE AND magic are similar according to Lewis is their encouragement of a stunning *lack* of skepticism. This may seem counterintuitive, since science in the popular imagination is supposed to be based on logic and evidence, while magic is supposed to be based on a superstitious acceptance of claims made in the name of the supernatural. In the words of Richard Dawkins, "[s]cience is based upon verifiable evidence," while "[r]eligious faith" (which Dawkins views as a kind of magic) "not only lacks evidence, its independence from evidence is its pride and joy."[25] Yet as Lewis well knew, scientific thinking no less than magical thinking can spawn a kind of credulity that accepts every kind of explanation no matter how poorly grounded in the facts. In the age of magic, the claims of the witch-doctor were accepted without contradiction. In the age of science, almost anything can be taken seriously if only it is defended in the name of science.

Lewis explained that one of the things he learned by giving talks at Royal Air Force camps during World War II was that the "real religion" of many ordinary Englishmen was a completely uncritical "faith in 'science.'"[26] Indeed, he was struck by how many of the men in his audiences "did not really believe that we have any reliable knowledge of historic man. But this was often curiously combined with a conviction that we knew a great deal about Pre-Historic Man: doubtless because Pre-Historic Man is labelled 'Science' (which is reliable) whereas Napoleon or Julius Caesar is labelled as 'History' (which is not)."[27]

But it was not just the "English Proletariat" who exhibited a credulous acceptance of claims made in the name of science according to Lewis. In *That Hideous Strength*, when the young sociologist Mark Studdock expresses doubts that N.I.C.E. can effectively propagandize "educated people," the head of N.I.C.E.'s police force, Fairy Hardcastle, responds tartly: "Why you fool, it's the educated reader who *can* be gulled. All our difficulty comes with the others. When did you meet a workman who believes the papers? He takes it for granted that they're all propaganda and skips the leading articles… We have to recondition him. But the educated public, the people who read the highbrow weeklies, don't need reconditioning. They're all right already. They'll believe anything."[28]

For Lewis, two leading examples of scientism-fueled gullibility of the intellectual classes during his own day were Freudianism and evolutionism.

Lewis's interest in Freud dated back to his days as a college student. In his *Surprised by Joy* (1955), he recalled how as an undergraduate "the new Psychology was at that time sweeping through us all. We did not swallow it whole… but we were all influenced."[29] In 1922 he recorded in his diary a discussion with friends saying that "[w]e talked a little of psychoanalysis, condemning Freud."[30] Although skeptical of Freud, Lewis remained intrigued, for a just few weeks later he notes that he was reading Freud's *Introductory Letters on Psychoanalysis*.[31]

A decade later, and shortly after Lewis had become a Christian, Freud made a cameo appearance in *The Pilgrim's Regress* (1933), Lewis's autobiographical allegory of his intellectual and spiritual journey toward Christianity.[32] In Lewis's story, the main character John ends up being arrested and flung into a dungeon by a stand-in for Freud named Sigismund Enlightenment (Sigismund was Sigmund Freud's full first name).[33] The dungeon is overseen by a Giant known as the Spirit of the Age who makes people transparent just by looking at them. As a result, wherever John turns, he sees through his fellow prisoners into their insides. Looking at a woman, he sees through her skull and "into the passages of the nose, and the larynx, and the saliva moving in the glands and the blood in the veins: and lower down the lungs panting like sponges, and the liver, and the intestines like a coil of snakes."[34] Looking at an old man, John sees the man's cancer growing inside him. And when John turns his head toward himself, he is horrified to observe the inner workings of his own body. After many days of such torment, John cries out in despair: "I am mad. I am dead. I am in hell for ever."[35]

The dungeon is the hell of materialistic reductionism, the attempt to reduce every human trait to an irrational basis, all in the name of modern science. Lewis saw Freud as one of the trailblazers of the reductionist approach. By attempting to uncover the "real" causes of people's religious and cultural beliefs in their subconscious and irrational urges and complexes, Freud eroded not only their humanity, but the authority of rational thought itself.

In the 1940s, Lewis offered an explicit critique of Freudianism in a lecture to the Socratic Club at Oxford. Noting that people used to believe that "if a thing seemed obviously true to a hundred men, then it was probably true in fact," Lewis observed that "[n]owadays the Freudian will tell you to go and analyze the hundred: you will find that they all think Elizabeth [I] a great queen because they all have a mother-complex. Their thoughts are psychologically tainted at the source."[36]

"Now this is obviously great fun," commented Lewis, "but it has not always been noticed that there is a bill to pay for it." If all beliefs are thus tainted at the source and so should be disregarded, then what about Freud's own system of belief? The Freudians "are in the same boat with all the rest of us... They have sawn off the branch they were sitting on."[37] In the name of a scientific study of psychology, the Freudians had undercut the confidence in reason needed for science itself to continue to flourish.[38]

Evolutionism was another prime example of credulous thinking fostered by scientism according to Lewis. As chapter 6 will explain, Lewis did not object in principle to an evolutionary process of common descent, although he was skeptical in practice of certain claims about common descent. But Lewis had no patience for the broader evolutionary idea that matter magically turned itself into complex and conscious living things through a blind and undirected process. Lewis lamented that "[t]he modern mind accepts as a formula for the universe in general the principle 'Almost nothing may be expected to turn into almost everything' without noticing that the parts of the universe under our direct observation tell a quite different story."[39] Fueled by "Darwinianism," this sort of credulity drew on "a number of false analogies" according to Lewis: "the oak coming from the acorn, the man from the spermatozoon, the modern steamship from the primitive coracle. The supplementary truth that every acorn was dropped by an oak, every spermatozoon derived from a man, and the first boat by something so much more complex than itself as a man of genius, is simply ignored."[40]

Lewis also thought that evolutionism, like Freudianism, promoted a "fatal self-contradiction" regarding the human mind.[41] According to the Darwinian view, "reason is simply the unforeseen and unintended by-product of a mindless process at one stage of its endless and aimless becoming." Lewis pointed out the fundamental difficulty with this claim: "If my own mind is a product of the irrational—if what seem my clearest reasonings are only the way in which a creature conditioned as I am is bound to feel—how shall I trust my mind when it tells me about Evolution?" He added that "[t]he fact that some people of scientific education cannot by any effort be taught to see the difficulty, confirms one's suspicion that we here touch a radical disease in their whole style of thought."[42]

Although science is supposed to be based on logic, evidence, and critical inquiry, Lewis understood that it could be easily misused to promote uncritical dogmatism, and he lived during an era in which this kind of misuse of science was rampant. Consider the burgeoning "science" of eugenics, the effort to breed better human beings by applying Darwinian principles of selection through imprisonment, forced sterilization, immigration restrictions, and other methods. Generally regarded today as pseudoscience, eugenics originated with noted British scientist Francis Galton (Charles Darwin's cousin), and it found widespread popularity in Lewis's day among elites in England, the United States, and Germany. Eugenics was the consensus view of the scientific community during much of Lewis's lifetime, and those who opposed it were derided as anti-science reactionaries or religious zealots standing in the way of progress. In America, its champions included members of the National Academy of Sciences and evolutionary biologists at the nation's top research universities.[43] In Britain, noted eugenists included evolutionary biologist Julian Huxley, grandson of "Darwin's Bulldog" Thomas Henry Huxley. Julian Huxley complained that in civilized societies "the elimination of defect by natural selection is largely… rendered inoperative by medicine, charity, and the social services." As a result, "[h]umanity will gradually destroy itself from within, will decay in its very core and essence, if this slow but relentless process is not checked."[44]

The United States holds the dubious honor of enacting the world's first compulsory eugenics sterilization law, but it was Nazi Germany that pursued eugenics with special rigor in the 1930s and 40s. Not content with merely sterilizing hundreds of thousands of the so-called "unfit," Nazi doctors eventually started killing handicapped persons en masse in what turned out to be a practice run for Hitler's extermination campaign against the Jews.[45]

The horrors of Nazi eugenics effectively killed off enthusiasm for eugenics in the mainstream scientific community after World War II. But there were other cases where scientific elites showed a similarly breathtaking lack of skepticism during this period. In the field of human evolution, much of the scientific community was hoodwinked for two generations into accepting the infamous Piltdown skull as a genuine "missing link" between humans and their ape-like ancestors before the fossil was definitively exposed as a forgery in 1953 (much to Lewis's private amusement).[46] In the field of medicine, the lobotomy was embraced as a miracle cure by large parts of the medical community well into the 1950s, and the scientist who pioneered the operation in human beings even won a Nobel Prize for his efforts in 1949. Only after tens of thousands of individuals had been lobotomized (including children) did healthy skepticism begin to prevail.[47] And in the field of human sexuality, Darwinian zoologist Alfred Kinsey's studies on human sex practices were accepted uncritically by fellow researchers and social scientists for decades despite the fact that his wildly unrepresentative samples and coercive interview techniques made his research little more than junk science.[48]

If scientists themselves could demonstrate such stunning bouts of credulity about scientific claims, members of the general public were even more susceptible to the disease according to Lewis. In an age of science and technology, Lewis knew that ordinary citizens must increasingly look to scientific experts for answers, and that would likely lead people to defer more and more to the scientists, letting the scientists do their thinking for them and neglecting their own responsibilities for critical thought in the process.

Lewis knew firsthand the dangers of simply deferring to scientific claims, recalling that his own atheistic "rationalism was inevitably based on what I believed to be the findings of the sciences, and those findings, not being a scientist, I had to take on trust—in fact, on authority."[49] Lewis understood that the ironic result of a society based on science might be greater credulity, not less, as more people simply accepted scientific claims on the basis of authority. This was already happening in his view. Near the end of his life, Lewis observed that "the ease with which a scientific theory assumes the dignity and rigidity of fact varies inversely with the individual's scientific education," which is why when interacting "with wholly uneducated audiences" he "sometimes found matter which real scientists would regard as highly speculative more firmly believed than many things within our real knowledge."[50] In Lewis's view, the increasing acquiescence of non-scientists to those with scientific and technical expertise gave rise to by far the most dangerous similarity between science and magic, one that threatened the future of Western civilization itself.

3. Science as Power

THE THIRD AND MOST SIGNIFICANT way science is similar to magic according to Lewis is its quest for power. Magic wasn't just about understanding the world; it was about controlling it. The great wizard or sorcerer sought power over nature. Similarly, science from the beginning was not just the effort to understand nature, but the effort to control it. "For magic and applied science alike the problem is how to subdue reality to the wishes of men," wrote Lewis. In pursuit of that objective, both magicians and scientists "are ready to do things hitherto regarded as disgusting and impious—such as digging up and mutilating the dead."[51]

Of course, there is a critically important difference between science and magic: Science works, while magic is relegated today to the pages of the fairy tale. Science cures diseases. Science increases food production. Science puts men on the moon and ordinary people in jet planes. Science fills our homes with computers, iPhones, and microwave ovens. Herein lies the great temptation of modern science to modern man. The world as we know it faces apparently insurmountable evils from hunger to disease to crime to war to ecological devastation. Science offers the hope of earthly salvation through the limitless creativity of human ingenuity—or so the prophets of scientism have claimed over the past century, including H. G. Wells and evolutionary biologists J. B. S. Haldane and Julian Huxley during C. S. Lewis's own day. Haldane viewed science as "man's gradual conquest, first of space and time, then of matter as such, then of his own body and those of other living beings, and finally the subjugation of the dark and evil elements in his own soul,"[52] and he urged his fellow scientists to no longer be "passively involved in the torrent of contemporary history, but actively engaged in changing society and shaping the world's future."[53]

C. S. Lewis was not persuaded. In his view, the scientific utopians failed to take into account the moral vacuum at the heart of contemporary science. Lewis stressed that he was not anti-science; but he still worried that modern science was ill-founded from the start: "It might be going too far to say that the modern scientific movement was tainted from its birth: but I think it would be true to say that it was born in an unhealthy neighbourhood and at an inauspicious hour."[54] Lewis noted that modern science attempts to conquer nature by demystifying its parts and reducing them to material formulas by which they can be controlled.

The results of this materialistic reductionism are often laudable (e.g., antibiotics, personal computers, and the invention of airplanes). Nevertheless, when the conquest of nature is turned on man himself, a problem arises: "[A]s soon as we take the final step of reducing our own species to the level of mere Nature, the whole process is stultified, for this time the being who stood to gain and the being who has been sacrificed are one and the same."[55] By treating human beings as the products of blind non-rational forces, scientific reductionism eliminates man as a rational moral agent. In Lewis's words, "[m]an's final conquest has proved to be the abolition of Man."[56]

Lewis worried that scientism's reductionist view of the human person would open the door wide to the scientific manipulation of human beings. "[I]f man chooses to treat himself as raw material," he wrote, "raw material he will be: not raw material to be manipulated, as he fondly imagined, by himself, but by mere appetite, that is, mere Nature, in the person of his dehumanized Conditioners."[57] Lewis thought there would be no effective limits on human manipulation in the scientific age because scientism undermined the authority of the very ethical principles needed to justify such limits. According to scientism, old cultural rules (such as "Man has no right to play God" or "punishment should be proportionate to the crime") were simply the byproducts of a blind evolutionary process and could be disregarded or superseded as needed. Thus, any restrictions on the application of science to human affairs ultimately would be left to the personal whims of the elites.

Lewis's concern about the powerful impact of scientism on society was detectible already in *Dymer* (1926) and *The Pilgrim's Regress* (1933), but by the late 1930s and early 1940s his alarm was on full display in his science fiction trilogy, which he continued to publish as the world plunged into another world war. It is significant that Lewis spent World War II writing not about the dangers of Nazism or communism (even though he detested both), but about the dangers of scientism and its effort to abolish man.[58] Scientism was a greater threat in Lewis's view than fascism or communism because it infected representative democracies like Britain no less than totalitarian societies: "The process which, if not checked, will abolish Man, goes on apace among Communists and Democrats no less than among Fascists." Lewis acknowledged that "[t]he methods may (at first) differ in brutality" between scientism and totalitarianism, but he went on to make a shocking claim: "[M]any a mild-eyed scientist in pince-nez, many a popular dramatist, many an amateur philosopher in our midst, means in the long run just the same as the Nazi rulers of Germany."[59]

That message lies at the heart of Lewis's novel *That Hideous Strength*, written in 1942 and 1943, but not published until 1945.[60] As previously mentioned, *That Hideous Strength* tells the story of a sinister conspiracy to turn England into a scientific utopia. The vehicle of transformation is to be a lavishly funded new government bureaucracy with the deceptively innocuous name of the National Institute for Co-ordinated Experiments, or N.I.C.E. for short.[61] Of course, there is nothing nice about N.I.C.E. Its totalitarian goal is to meld the methods of modern science with the coercive powers of government in order "to take control of our own destiny" and "make man a really efficient animal." The Institute's all-encompassing agenda reads like a wish list drawn up by the era's leading scientific utopians: "sterilization of the unfit, liquidation of backward races (we don't want any dead weights), selective breeding," and "real education," which means "biochemical conditioning... and direct manipulation of the brain."[62] N.I.C.E.'s agenda also includes scientific experimentation on both animals and criminals. The animals would be "cut up like paper on the mere chance of some interesting discovery," while the criminals would no longer be punished but cured, even if their "remedial treatment" must continue indefinitely.[63]

Lewis lampoons the scientific bureaucrats running N.I.C.E., and he relishes pointing out just how narrow-minded and parochial they are for all of their supposed sophistication. This comes out clearly when Mark Studdock and a fellow researcher from the sociology branch of N.I.C.E. (Cosser) visit a picturesque country village in order to write a report advocating its demolition. Mark, who is not quite as far down the path of scientism as Cosser, feels like he is "on a holiday" while visiting the village, enjoying the natural beauty of the sunny winter day, relaxing at a pub for a drink, and feeling the aesthetic attraction of historic English architecture. Cosser is impervious to such things, placing no value on anything outside his narrow field of sociological expertise. Instead of delighting in the beauty of nature, Cosser complains about the "[b]loody awful noise those birds make."[64] Instead of enjoying a drink at the pub, he complains about the lack of ventilation and suggests that the alcohol could be "administered in a more hygienic way." When Mark suggests that Cosser is missing the point of the pub as a gathering place for food and fellowship, Cosser replies "Don't know, I'm sure… Nutrition isn't my subject. You'd want to ask Stock about that." When Mark mentions that the village has "its pleasant side" and that they need to make sure that whatever it is replaced with is something better in all areas, "not merely in efficiency," Cosser again pleads that this is outside his area. "Oh, architecture and all that," he replies. "Well, that's hardly my line, you know. That's more for someone like Wither. Have you nearly finished?"[65] A hyper-specialist, Cosser can't see past his proverbial nose. Yet he is being given the power to decide whether to dispossess members of an entire village from their homes.[66]

That Hideous Strength resonated with the public, and it quickly became Lewis's most popular adult novel, despite negative reviews from critics, including one from J. B. S. Haldane, who thought the novel was a blatant attack on science.[67] It is easy to understand why the public of the 1940s might have been receptive to the novel's message. Two world wars and the rise of totalitarianism in Germany and Russia had dampened popular enthusiasm for the message of the scientific utopians. After all, it was hard to view science as savior when scientists were busy bringing forth poison gas, the V-2 rocket, and the atomic bomb—not to mention new methods of killing the handicapped in the name of eugenics in Germany. To many people, the new age ushered in by science looked more like a nightmare than a paradise.

After World War II, however, even the looming threat of nuclear annihilation did not prevent some from renewing their quest for societal salvation through science, and scientific utopianism began to revive. At the global level, Julian Huxley called for bringing about a better future by promoting "scientific world humanism, global in extent and evolutionary in background,"[68] while in America renewed optimism toward science was exemplified by icons of pop culture such as Walt Disney's "Tomorrowland" in Disneyland, *The Jetsons* cartoon series, and the 1962 World's Fair in Seattle, which celebrated the seemingly endless possibilities of the science-led world of "Century 21."

For his part, Lewis continued to sound the alarm about the dangers of what he variously called "technocracy" or "scientocracy"—government in the name of science that is disconnected from the traditional limits of both morality and a free society.[69] Lewis's most eloquent post-war statement on the subject came in the article "Willing Slaves of the Welfare State," published in *The Observer* in 1958. In that essay, Lewis worried that we were seeing the rise of a "new oligarchy [that] must more and more base its claim to plan us on its claim to knowledge… This means they must increasingly rely on the advice of scientists, till in the end the politicians proper become merely the scientists' puppets."[70] Lewis believed that the world's desperate ills of "hunger, sickness, and the dread of war" would make people all too willing to accept an "omnicompetent global technocracy," even if it meant surrendering their freedoms. "Here is a witch-doctor who can save us from the sorcerers—a war-lord who can save us from the barbarians—a Church that can save us from Hell. Give them what they ask, give ourselves to them bound and blindfold, if only they will!"[71]

Lewis did not deny that scientific and technical knowledge might be needed to solve our current problems. But he challenged the claim that scientists had the right to rule merely because of their superior technical expertise. Scientific knowledge may be necessary for good public policy in certain areas, but Lewis knew that it was hardly sufficient. Political problems are preeminently moral problems, and scientists are ill-equipped to function as moralists according to Lewis: "Let scientists tell us about sciences. But government involves questions about the good for man, and justice, and what things are worth having at what price; and on these a scientific training gives a man's opinion no added value."[72]

Lewis's warnings about the threat of scientocracy could have come from the latest headlines. Since the 1990s there has been a dramatic increase in what some have called the "authoritarian tone" of science, exemplified by the growing use in science journalism during this period of phrases such as "science requires," "science dictates," and "science tells us we should."[73] The changes in journalism track with similar developments in politics and public policy. Whether the topic be embryonic stem cell research, climate change, health insurance mandates, the teaching of evolution, or any number of other topics, "science" is increasingly being used as a trump card in public debates to suppress dissent and curtail discussion. Regardless of the issue, experts assert that their public policy positions are dictated by "science," which means that anyone who disagrees with them is "anti-science."

The conflict over government funding for embryonic stem cell research is a perfect example. Oppose taxpayer funding for embryonic stem cell research, and you are guaranteed to be labeled "anti-science" as well as a religious fanatic. However, this storyline of enlightened scientists vs. intolerant fundamentalists opposed to research obscures the complexities of the actual debate. First, there are plenty of scientific (as opposed to ethical or religious) objections to the efficacy of embryonic stem cell research; these are conveniently ignored by framing the dispute as science vs. anti-science.[74] Second, raising ethical questions about certain kinds of scientific research makes one "anti-science" only if one accepts scientism's premise that science is the one valid form of knowledge in the public square and scientific research therefore should operate free from any outside restrictions whatever. According to this mindset, opposition to the infamous Tuskegee syphilis experiments or Nazi medical experimentation on Jews would make one "anti-science." But that is ridiculous. Practicing science does not require operating in a moral vacuum, and raising ethical objections to some forms of scientific research does not make one "anti-science."

A similar situation exists in the debate over climate change. Question any part of the climate change "consensus" (how much climate change is going on, how much humans contribute to it, or what should humans do about it), and one is instantly declared "anti-science" or even a threat to the future of the human race. The goal of this kind of rhetoric is not to win by persuading others, but by silencing them.

Along with the growing use of science as a trump card, we are seeing the revival of scientific justifications for eugenics under the banners of "Transhumanism" (see chapter 10) and "reprogenetics." The latter term was coined by Princeton University biologist Lee Silver, who urges human beings to take control of their evolution and evolve themselves into a higher race of beings with god-like powers.[75] Although Silver is concerned that the supposed blessings of genetic engineering might not be equally distributed across the population,[76] he nonetheless urges us to seize the opportunity: "[H]uman beings... now have the power not only to control but to create new genes for themselves. Why not seize this power? Why not control what has been left to chance in the past?"[77] "Transhumanism" and "reprogenetics" may still sound like science fiction to many people, but eugenic abortions targeting children with genetic defects are already well under way. In 2012, physician Nancy Synderman, chief medical editor for NBC News, publicly defended eugenic abortions on national television squarely on the basis of science: "I am pro-science, so I believe that this is a great way to prevent diseases."[78] Of course, if it is "pro-science" to support eradicating babies with genetic flaws, it must be "anti-science" to oppose it.

For the moment, the new eugenics is focused more on encouraging individuals to willingly breed a better race than on imposing top-down measures, but the use of science as a justification for coercion is on the upswing as well:

- In the name of saving the planet from global warming, British scientist James Lovelock has called for the suspension of democracy: "Even the best democracies agree that when a major war approaches, democracy must be put on hold for the time being. I have a feeling that climate change may be an issue as severe as a war. It may be necessary to put democracy on hold for a while."[79]

- In the name of promoting biodiversity, evolutionary zoologist Eric Pianka at the University of Texas urges the reduction of the Earth's human population by up to 90% and calls on the government to confiscate all the earnings of any couple who have more than two children. "You should have to pay more when you have your first kid—you pay more taxes," he insists. "When you have your second kid you pay a lot more taxes, and when you have your third kid you don't get anything back, they take it all."[80]

- In order to achieve the admittedly laudable goal of ending obesity, Harvard evolutionary biologist Daniel Lieberman advocates coercive measures by the government to control our diets. Lieberman argues that coercion is necessary because evolutionary biology shows us that we cannot control our sugar intake on our own power. "We have evolved to need coercion."[81]

- When the Obama administration mandated that many private religious employers include contraceptives and even certain kinds of abortion drugs as part of their health care plans, the abrogation of religious liberty rights was justified in the name of science. "Scientists have abundant evidence that birth control has significant health benefits for women," declared Secretary of Health and Human Services Kathleen Sebelius, defending the mandate.[82]

Lewis's age of scientocracy has come upon us with a vengeance. Now we need to figure out what to do about it.

A Regenerate Science?

Lewis provides a hint as to what will be required to overcome scientism in his Narnian story *The Magician's Nephew*. Despite its title, there are actually two magicians in the story. The first, Uncle Andrew, embodies the longing to fuse science with magic. Although a magician, Uncle Andrew is also a scientist. He has a microscope, and he experiments on animals.[83] By pursuing power over nature without regard to ethics, Uncle Andrew sets in motion a train of events that ultimately brings a far greater magician, Queen Jadis, into both Earth and Narnia, which she thereupon threatens to enslave. Jadis previously destroyed her own world, Charn, after using her knowledge of "the Deplorable Word" to liquidate the entire population of the planet. The "Deplorable Word" was a secret formula "which, if spoken with the proper ceremonies, would destroy all living things except the one who spoke it." Previous rulers of Charn had pledged never to seek knowledge of the formula, but Jadis violated her oath, and when faced with defeat in battle, she decided to use the word.[84]

Jadis is ultimately thwarted in her effort to take over new worlds, not by the actions of a fellow magician, but by the repentance of a young boy, Digory. Digory's unconstrained curiosity previously had brought Jadis out of a deep sleep. In order to undo the harm brought about by awakening Jadis, Digory promises Aslan, the Creator of Narnia, that he will journey to a garden on top of the mountains where he will pick a magical apple and bring it back to Aslan. When Digory arrives at the garden, he finds Jadis already there, having gorged herself on one of the apples despite a sign forbidding people to take apples for themselves. Jadis then urges Digory to disregard his promise to Aslan and take an apple for his dying mother, assuring him that the apple will heal her of her illness. Even when Jadis accuses Digory of being "heartless" for not being willing to save his own mother, Digory rebuffs the temptation to break faith with Aslan. As a result of Digory's unwillingness to cooperate with her evil scheme, Jadis and her evil power are kept in check for many centuries.[85]

The Magician's Nephew was written during the 1950s, the very period when Lewis's concerns about an "omnicompetent global technocracy" continued to grow. Jadis clearly represents the dangers of scientism. Her use of the "Deplorable Word" in her own world is perhaps a commentary on the age of nuclear weapons and our own efforts to develop ever more powerful weapons of mass destruction. After Aslan says that humans should take warning from the destruction of Charn, Digory's friend Polly says: "But we're not quite as bad as that world, are we, Aslan?" Aslan responds: "Not yet. But you are growing more like it. It is not certain that some wicked one of your race will not find out a secret as evil as the Deplorable Word and use it to destroy all living things." Aslan then tells Digory and Polly that "before you are an old man and an old woman, great nations in your world will be ruled by tyrants who care no more for joy and justice and mercy than the Empress Jadis. Let your world beware."[86] Since *The Magician's Nephew* is set in the early 1900s, Aslan is undoubtedly referring to the two world wars and subsequent "Cold War" that loomed on the horizon, all of which would be accompanied by horrifying new uses of science and technology to kill and manipulate humanity.[87]

In *The Abolition of Man*, Lewis expressed his hope that a reformation of science could be brought about by scientists. But he made clear that the task was too important to be left to them alone: "[I]f the scientists themselves cannot arrest this process before it reaches the common Reason and kills that too, then someone else must arrest it."[88] In a free society, scientism requires the cooperation of scientists and non-scientists alike to prevail, and it requires the cooperation of both scientists and non-scientists to be defeated.

Like Digory, people today need the courage and independence of thought to stand up to the magicians of scientism. They need to be willing to ask questions, challenge assumptions, and defend a broader view of rationality than that permitted by scientific materialism. Whether the issue is climate change, embryonic stem cell research, genetic engineering, evolution and intelligent design, or something else, it is not enough to simply acquiesce in the current "climate of opinion" in science or anything else, as Lewis himself well knew. "I take a very low view of 'climates of opinion,'" he commented, noting that "[i]n his own subject every man knows that all discoveries are made and all errors corrected by those who *ignore* the 'climate of opinion.'"[89]

At the end of *The Abolition of Man*, Lewis issued a call for a "regenerate science" that would seek to understand human beings and other living things as they really are, not try to reduce them to automatons. "When it explained it would not explain away. When it spoke of the parts it would remember the whole. While studying the *It* it would not lose what Martin Buber calls the *Thou*-situation."[90]

Lewis was not quite sure what he was asking for, and he was even less sure that it could come to pass. Yet in recent decades we have begun to see glimmers. New developments in biology, physics, and cognitive science are raising serious doubts about the most fundamental tenets of scientific materialism. In physics, our understanding of matter itself is becoming increasingly non-material.[91] In biology, scientists are discovering how irreducibly complex biological systems and information encoded in DNA are pointing to the reality of intelligent design in nature.[92] In cognitive science, efforts to reduce mind to the physical processes of the brain continue to fail, and new research is providing evidence that the mind is a non-reducible reality that must be accepted on its own terms.[93] What George Gilder has called "the materialist superstition" is being challenged as never before.[94]

Nearly 50 years after C. S. Lewis's death, we are facing the possibility that science can become something more than the magician's twin. Even in the face of surging scientism in the public arena, an opportunity has opened to challenge scientism on the basis of science itself, fulfilling Lewis's own desire that "from Science herself the cure might come."[95]

Let us hope we find the clarity and courage to make the most of the opportunity.

Endnotes

1. C. S. Lewis, *The Abolition of Man* (New York: Macmillan, 1955), 87.
2. C. S. Lewis, *The Magician's Nephew* (New York: Macmillan, 1955), and *That Hideous Strength* (New York: Macmillan, 1965).
3. C. S. Lewis, "Is Theology Poetry?" in *The Weight of Glory and Other Addresses*, revised and expanded edition (New York: Macmillan, 1980), 79; C. S. Lewis, "The Funeral of a Great Myth," in *Christian Reflections*, edited by Walter Hooper (Grand Rapids: Eerdmans, 1967), 82–93; C. S. Lewis, "The World's Last Night," in *The World's Last Night and Other Essays* (San Diego: Harcourt Brace & Company, 1960), 101.
4. Tom Bartlett, "Is Evolution a Lousy Story?," *The Chronicle of Higher Education*, May 2, 2012, accessed June 12, 2012, http://chronicle.com/blogs/percolator/is-evolution-a-lousy-story/29158?sid=pm&utm_source=pm&utm_medium=en.
5. Lewis, "Is Theology Poetry?" 79.
6. H. G. Wells, *A Short History of the World* (New York: Macmillan, 1922), 16, accessed June 13, 2012, http://www.gutenberg.org/files/35461/35461-h/35461-h.htm.
7. Ibid., 426–427.
8. H. G. Wells, *The Time Machine* (1898), chapter 12, accessed June 13, 2012, http://www.gutenberg.org/cache/epub/35/pg35.html.
9. Lewis, "Funeral of a Great Myth," 88.
10. C. S. Lewis, *Miracles: A Preliminary Study* (New York: Macmillan, 1960), 106.
11. Cathy Lynn Grossman, "Richard Dawkins to Atheist Rally: 'Show Contempt' for Faith," *USA Today*, March 24, 2012, accessed June 12, 2012, http://content.usatoday.com/communities/Religion/post/2012/03/-atheists-richard-dawkins-reason-rally/1#.T8AiFb-iyJg; Kimberly Winston, "Atheists Rally on National Mall," *The Huffington Post*, March 24, 2012, accessed June 12, 2012, http://www.huffingtonpost.com/2012/03/24/atheist-rally_n_1377443.html.
12. Richard Dawkins, *The Blind Watchmaker: Why the Evidence of Evolution Reveals a Universe Without Design* (New York: W.W. Norton and Co., 1996), 6; Michael Shermer, "Science Is My Savior," accessed Feb. 23, 2007, http://www.science-spirit.org/article_detail.php?article_id=520. Shermer's article appears to be no longer available online.
13. "About the International Darwin Day Foundation," International Darwin Day Foundation website, accessed June 12, 2012, http://darwinday.org/about-us/.
14. Quoted in John G. West, *Darwin Day in America: How Our Politics and Culture Have Been Dehumanized in the Name of Science* (Wilmington: ISI Books, 2007), 210.
15. Comment by "Evolution171," International Darwin Day Foundation website, accessed June 12, 2012, http://darwinday.org/about/#comment-444861134.
16. "America's Evolutionary Evangelists," accessed June 12, 2012, http://evolutionaryevangelists.libsyn.com/.
17. Michael Dowd, *Thank God for Evolution: How the Marriage of Science and Religion Will Transform Your Life and Our World* (New York: Plume, 2009).

18. Description at The Great Story website, accessed June 12, 2012, http://evolutionaryevangelists.libsyn.com/. Emphasis in the original.
19. See, in particular, the links on the home page for "Great Story Parables," "Great Story Beads," "Songs," "Children's Curricula," and "Group Study Support Materials," The Great Story website, accessed June 12, 2012, http://www.thegreatstory.org/.
20. Michael Dowd, "Thank God for the New Atheists," *Skeptic Magazine*, 16, no. 2 (2011): 29.
21. Ibid., 29-30.
22. Comments by Loyal Rue, highlighted on The Great Story website, accessed June 12, 2012, http://www.thegreatstory.org/what_is.html. Emphasis added.
23. Phillip Johnson, "In the Beginning Were the Particles," Grace Valley Christian Center, March 5, 2000, accessed June 12, 2012, http://gracevalley.org/sermon_trans/Special_Speakers/In_Beginning_Particles.html.
24. Lewis, *That Hideous Strength*, 79.
25. Richard Dawkins, "Is Science a Religion?" *The Humanist* (January/February 1997), accessed June 12, 2012, http://www.thehumanist.org/humanist/articles/dawkins.html.
26. C. S. Lewis, "Christian Apologetics," *God in the Dock*, edited by Walter Hooper (Grand Rapids: Eerdmans, 1970), 95.
27. C. S. Lewis, "God in the Dock," *God in the Dock*, 241.
28. Lewis, *That Hideous Strength*, 99–100.
29. C. S. Lewis, *Surprised by Joy* (New York: Harcourt Brace Jovanovich, 1955), 203.
30. Entry for Thursday, May 25, 1922, C. S. Lewis, *All My Road Before Me: The Diary of C. S. Lewis, 1922–1927*, edited by Walter Hooper (San Diego: Harcourt Brace Jovanovich, 1991), 41.
31. Ibid., 44.
32. C. S. Lewis, *The Pilgrim's Regress* (New York: Bantam Books, 1981).
33. Armand M. Nicholi, Jr., *The Question of God: C. S. Lewis and Sigmund Freud Debate God, Love, Sex, and the Meaning of Life* (New York: Free Press, 2002), 5; Kathryn Lindskoog, *Finding the Landlord: A Guidebook to C. S. Lewis's* Pilgrim's Regress (Chicago: Cornerstone Press, 1995), 31–33.
34. Ibid., 48.
35. Ibid., 49.
36. C. S. Lewis, "Bulverism," *God in the Dock*, 271.
37. Ibid., 272.
38. Despite Lewis's withering critiques of Freud, he did not reject psychoanalysis out of hand, nor the obvious truth that many of our beliefs may be influenced by non-rational factors. See C. S. Lewis, *Mere Christianity* (New York: Macmillan, 1960), 83–84; C. S. Lewis, *Letters to Malcolm: Chiefly on Prayer* (San Diego: Harcourt Brace Jovanovich, 1964), 34.
39. Ibid., 64.
40. Ibid., 65–66.
41. Lewis, "Funeral of a Great Myth," 88.

42. Ibid., 89.
43. West, *Darwin Day in America*, 86–88, 122–162.
44. Julian Huxley, "Eugenics and Society," *Eugenics Review*, 28, no. 1 (1936): 30–31, accessed June 12, 2012, http://www.thegreatstory.org/what_is.html.
45. See Leo Alexander, "Medical Science under Dictatorship," *New England Journal of Medicine*, 241, no. 2 (1949): 39–47, accessed June 13, 2012, doi:10.1056/NEJM194907142410201.
46. For more information about the Piltdown hoax, see Frank Spencer, *Piltdown: A Scientific Forgery* (New York: Oxford University Press, 1990). For Lewis's reaction to the exposure of Piltdown, see the discussion in chapter 6 of the present book.
47. West, *Darwin Day in America*, 88–92.
48. Ibid., 268–290.
49. Lewis, *Surprised by Joy*, 174.
50. C. S. Lewis, *The Discarded Image* (Cambridge: Cambridge University Press, 1964), 17.
51. Lewis, *Abolition of Man*, 88.
52. J.B.S. Haldane, Daedalus, or, Science and the Future (1923), accessed June 13, 2012, http://www.marxists.org/archive/haldane/works/1920s/daedalus.htm.
53. J.B.S. Haldane, "Dialectical Materialism and Modern Science: IV. Negation of the Negation," Labour Monthly (October 1941): 430–432, accessed June 12, 2012, http://www.marxists.org/archive/haldane/works/1940s/dialectics04.htm.
54. Lewis, *Abolition of Man*, 89.
55. Ibid., 83.
56. Ibid., 77.
57. Ibid., 84.
58. For a discussion of Lewis's critique of both communism and fascism, see John G. West, "Communism and Fascism," in Jeffrey D. Schultz and John G. West, *The C. S. Lewis Readers' Encyclopedia* (Grand Rapids: Zondervan, 1998), 126–127.
59. Ibid., 85.
60. Walter Hooper, *C. S. Lewis: A Companion & Guide* (San Francisco: HarperSanFrancisco, 1996), 231–232.
61. Ironically, the British government in the 1990s actually created a controversial government bureaucracy with the same acronym to ration health care. See the website of the National Institute for Health and Clinical Excellence (NICE), accessed June 12, 2012, http://www.nice.org.uk/. For criticism of the real NICE see "Of NICE and Men," *The Wall Street Journal*, July 7, 2009, accessed June 13, 2012, http://online.wsj.com/article/SB124692973435303415.html.
62. Lewis, *That Hideous Strength*, 42.
63. Ibid., 102, 69.
64. Ibid., 87.
65. Ibid., 88.
66. Ibid., 85–88.

67. George Sayer, *Jack: A Life of C. S. Lewis* (Wheaton: Crossway, 1995), 304. For J.B.S. Haldane's negative review of *That Hideous Strength*, see Haldane, "Auld Hornie, F.R.S.," *Modern Quarterly* (Autumn 1946), accessed June 14, 2012, http://www.lewisiana.nl/haldane/#Auld_Hornie. For Lewis's reply, see "Reply to Professor Haldane," in *Of Other Worlds: Essays and Stories*, ed. Walter Hooper (New York: Harcourt Brace Jovanovich, 1966), 74–85.
68. Julian Huxley, UNESCO: Its Purpose and Its Philosophy (Preparatory Commission of the United Nations Educational, Scientific and Cultural Organisation, 1946), 8.
69. C. S. Lewis to Dan Tucker, Dec. 8, 1959, in *The Collected Letters of C. S. Lewis*, edited by Walter Hooper (San Francisco: HarperSanFrancisco, 2007), vol. III, 1104; Lewis, "Is Progress Possible? Willing Slaves of the Welfare State," *God in the Dock*, 311–316.
70. Lewis, "Is Progress Possible? Willing Slaves of the Welfare State," 314.
71. Ibid., 316.
72. Ibid., 315.
73. Kenneth P. Green and Hiwa Alaghebandian, "Science Turns Authoritarian" (July 27, 2010), accessed June 12, 2012, http://www.american.com/archive/2010/july/science-turns-authoritarian.
74. David Klinghoffer, "The Stem Cell War," *National Review Online*, April 13, 2011, accessed June 13, 2012, http://www.nationalreview.com/articles/264551/stem-cell-war-david-klinghoffer#; Wesley J. Smith, "The Wrong Tree," *National Review Online*, May 13, 2004, accessed June 13, 2012, http://www.discovery.org/a/2039; David A. Prentice, "Current Science of Regenerative Medicine with Stem Cells," *Journal of Investigative Medicine*, 54, no. 1 (January 2006): 33–37.
75. Lee Silver, *Remaking Eden* (New York: Harper Perennial, 1998).
76. Ibid., 13.
77. Ibid., 277.
78. Quoted in Kyle Drennen, "NBC: It's 'Pro-Science' to Abort Children with Genetic Defects," June 12, 2012, accessed June 12, 2012, http://www.lifenews.com/2012/06/12/nbc-its-pro-science-to-abort-children-with-genetic-defects/.
79. Quoted in Leo Hickman, "James Lovelock: Humans are too stupid to prevent climate change," March 29, 2010, accessed June 12, 2012, http://www.guardian.co.uk/science/2010/mar/29/james-lovelock-climate-change.
80. Eric R. Pianka, "The Vanishing Book of Life on Earth," accessed June 13, 2012, http://www.zo.utexas.edu/courses/bio373/Vanishing.Book.pdf; Jamie Mobley, "Doomsday: UT prof says death is imminent," *Seguin Gazette-Enterprise*, Feb. 27, 2010.
81. Daniel Lieberman, "Evolution's Sweet Tooth," *The New York Times*, June 5, 2012, accessed June 13, 2012, http://www.nytimes.com/2012/06/06/opinion/evolutions-sweet-tooth.html.
82. Robert Pear, "Obama Reaffirms Insurers Must Cover Contraception," *The New York Times*, Jan. 20, 2012, accessed June 12, 2012, http://www.nytimes.com/2012/01/21/health/policy/administration-rules-insurers-must-cover-contraceptives.html.
83. Lewis, *Magician's Nephew*, 9, 11, 19.
84. Ibid., 44–55.

85. Ibid., 14–149.
86. Ibid., 159–160.
87. Chapter 1 of *The Magician's Nephew* says it was set when "the Bastables were looking for treasure in Lewisham Road." Ibid., 1. The Bastables were characters in stories published by Edith Nesbit starting in 1899.
88. Lewis, *Abolition of Man*, 90.
89. Lewis, *The Problem of Pain* (New York: Macmillan, 1962), 134.
90. Lewis, *Abolition of Man*, 89–90.
91. See discussion in West, *Darwin Day in America*, 373–374.
92. Ibid., 370–373. Also see Michael Behe, *Darwin's Black Box: The Biochemical Challenge to Evolution*, second edition (New York: Free Press, 2006); Stephen Meyer, *Signature in the Cell: DNA and the Evidence for Intelligent Design* (New York: HarperOne, 2009); Bruce L. Gordon and William A. Dembski, editors, *The Nature of Nature* (Wilmington: ISI Books, 2011).
93. West, *Darwin Day in America*, 274–375; Mario Beauregard and Denyse O'Leary, *The Spiritual Brain* (New York: HarperOne, 2007).
94. George Gilder, "The Materialist Superstition," *The Intercollegiate Review*, 31, No. 2 (Spring 1996): 6–14.
95. Lewis, *Abolition of Man*, 87.

2

C. S. Lewis on Mere Science

M. D. Aeschliman

In *The Abolition of Man*, C. S. Lewis noted that nothing he could say would keep some people from saying that he was anti-science, a charge he was nevertheless eager to refute.[1] In fact, he had received the kind of philosophical education at Oxford that enabled him, like John Henry Newman before him, to appreciate what might be described as "mere science" while resisting the two opposed temptations that the historian of science Richard Olson has labeled "science deified" and "science defied."[2]

"Science deified" is scientism, radical empiricism, materialism, or naturalism, an implicit or explicit rejection of all nonquantifiable realities or truths, including the truths of reason. Its logical terminus is determinism or "epiphenomenalism," T. H. Huxley's notion that the brain and mind are fully determined by-products of irrational physical processes. As the German materialist Karl Vogt put it, "Thoughts come out of the brain as gall from the liver, or urine from the kidneys," implying that thoughts are just as irrational and beyond our control.[3] Vogt and the other materialists contradict themselves, though, because—as Lewis often noted—they claim that their own scientific thoughts are true.

The deification of science first became explicit in the writings of the atheistic French philosophes La Mettrie, D'Holbach, and Diderot. Thoughtful twentieth-century commentators such as Lester G. Crocker and Aldous Huxley have seen its reductionism leading straight to the moral nihilism of the Marquis de Sade, and later to Social Darwinism and the Nietzschean transvaluation of values in the interest of amoral strength and force.[4] Lewis's *Abolition of Man* is, among other things, an extended treatise against the deification of science.

Yet there is an opposite temptation that Lewis also criticized—the temptation to defy science, from the standpoint of either romantic/pantheistic gnosticism or theological fideism. The first was familiar to him from the theosophy of his close friends Owen Barfield and A. C. Harwood and from the whole history of Romanticism, culminating in the work and world of W. B. Yeats. (Yeats was probably the model for the magician in Lewis's *Dymer* and for Merlin in *That Hideous Strength*.[5]) The appeal of pantheistic gnosticism was something that Lewis understood and withstood; it lies at the heart of occult "New Age" spirituality, "Deep Ecology," and a good deal of "Eco-feminism" today.[6] Romantic self-absorption and pantheistic gnosticism are targets of Lewis's satire in *The Pilgrim's Regress*.[7] Much as he criticized radical empiricism and its sterile, truncated rationalism, he was himself too much of a rationalist in the classic, Aristotelian sense to countenance esoteric or occult mysticism and the depreciation of reason. He would not defy science on romantic or gnostic grounds.

Lewis knew that science was one of the great products of the human mind, but he insisted that it was a subset of reason and not simply equivalent to it. Scientific reason, if accurate, was valid, but it was not the only valid kind of reasoning: Noncontradiction, validity, truth, value, meaning, purpose, and obligation were necessary presuppositions of the scientific method but not themselves scientific phenomena. Lewis thought that, in Alfred North Whitehead's words, scientists who were "animated by the purpose of proving that they are purposeless constitute an interesting subject for study."[8] He satirically depicted such scientists in *That Hideous Strength*, especially in the figure of Frost. Of all radical empiricists, from La Mettrie and Hume to A. J. Ayer, who would undermine the authority of reason and its procedures, Lewis tirelessly pointed out this contradiction. He believed in E. A. Burtt's old adage that "the only way to avoid metaphysics is to say nothing,"[9] because in some important sense language and thought themselves are non-natural, supernatural, transcendent, and metaphysical. "In order to think," he wrote in 1942, "we must claim for our reasoning a validity which is not credible if our own thought is merely a function of our brain, and our brains a by-product of irrational physical processes."[10]

Lewis's love of the Middle Ages and the Renaissance was due largely to his loyalty to an epistemology that he thought had been caricatured and misunderstood by Bacon, Descartes, and the French Encyclopedists of the eighteenth century. As a careful student of the history of philosophy and ideas, he knew that the great flowering of scientific thought in the seventeenth century had not only Greek roots, but medieval ones. Whitehead pointed out long ago, in *Science and the Modern World*, that the habits of medieval rationalism prepared the way for the scientific discoveries of the seventeenth century, an insight given far more documentation, depth, and scope in the writings of the historian and philosopher of science Stanley L. Jaki in our time.[11] Long before Bacon, Jaki has written, Christian philosophy had steadily inculcated "the conviction... that since the world was rational it could be comprehended by the human mind, but as the product of the Creator it could not be derived from the mind of man, a creature."[12] The "metaphysical realism" of St. Thomas Aquinas (and of Richard Hooker in England) avoided the extremes of empiricism and idealism and thus paved the way for Newton.

Jaki's work has confirmed some of Lewis's insights about the origin and development of Western science, and particularly its indebtedness to the doctrine of creation *ex nihilo* to escape from mistaken Aristotelian ideas about time and matter. The importance of the medieval thinkers Buridan and Oresme for science had been rediscovered by the great twentieth-century French physicist Pierre Duhem, whose own work Jaki has done so much to restore to the prominence it deserves. The active intellectual discrimination against Duhem, and subsequently against Jaki—despite their enormous erudition and unquestionable distinction—would not have surprised the man who wrote "The Inner Ring," "Bulverism," *The Abolition of Man*, and *That Hideous Strength*.[13]

Among historians of science it is most prominently Duhem and Jaki who have provided the documentation of the importance of theism and "metaphysical realism" not only for the origin and development of modern science, but also for the possibility of its coherent continuation and moral direction. Duhem and Jaki have provided security for Lewis's claim that:

> Men became scientific because they expected Law in Nature, and they expected Law in Nature because they believed in a Legislator. In most modern scientists this belief has died: it will be interesting to see how long their confidence in uniformity survives it. Two significant developments have already appeared—the hypothesis of a lawless sub-nature, and the surrender of the claim that science is true. We may be living nearer than we suppose to the end of the Scientific Age.[14]

And as a believer in the essential sanity and continuity of Western Civilization, Lewis would surely have concurred with Jaki's characterization of the Middle Ages:

> In Western history that was the first and thus far the last major epoch in which broadly shared respect was paid to the fundamental difference between ends and means... If we do not wish to help turn this most scientific age of ours into the most barbaric of all ages, we had better stop using the term "medieval" as synonymous with obscurantist. In doing so, we may make our mental eyes more sensitive to that light which comes from the Middle Ages.[15]

Endnotes

1. C. S. Lewis, *The Abolition of Man* (New York: Touchstone, 1996), 82.
2. Richard Olson, *Science Deified and Science Defied* (Los Angeles: University of California Press, 1982).
3. Vogt, quoted in Owen Chadwick, *The Secularization of the European Mind in the Nineteenth Century* (Cambridge: Cambridge University Press, 1975), 166.
4. See Lester G. Crocker, *Nature and Culture: Ethical Thought in the French Enlightenment* (Baltimore: Johns Hopkins Press, 1963), Aldous Huxley, *Ends and Means* (London: Chatto & Windus, 1937).
5. C. S. Lewis, *Dymer* (New York: Macmillan, 1950) and C. S. Lewis, *That Hideous Strength* (New York: Macmillan, 1946).
6. Theodore Roszak, "The Monster and the Titan: Science, Knowledge, and Gnosis," *Daedalus*, 103, No. 3 (1974): 17–32. Carolyn Merchant, *The Death of*

Nature: Women, Ecology, and the Scientific Revolution (New York: HarperOne, 1990).
7. C. S. Lewis, *The Pilgrim's Regress* (New York: Sheed and Ward, 1935).
8. Whitehead, quoted by Joseph A. Mazzeo, *The Theory and Practice of Interpretation: A Commonplace Book* (New York: privately printed, 1975), 65.
9. Quoted by Stanley L. Jaki, "The Demythologization of Science," in *"The Absolute beneath the Relative" and Other Essays* (Lanham, MD: University Press of America, 1988), 201.
10. C. S. Lewis, "Miracles," in Walter Hooper, editor, *God in the Dock: Essays on Theology and Ethics* (Grand Rapids, MI: Eerdmans, 1970), 27.
11. Alfred North Whitehead, *Science and the Modern World* (New York: Free Press, 1997).
12. Stanley L. Jaki, *The Origin of Science and the Science of Its Origin* (Edinburgh: Scottish Academic Press, 1978), 21.
13. See C. S. Lewis, "Bulverism" in *God in the Dock*, pp. 271–277, and C. S. Lewis, "The Inner Ring," in *The Weight of Glory and Other Addresses*, revised and expanded edition (New York: Macmillan, 1980), 93–105.
14. C. S. Lewis, *Miracles: A Preliminary Study* (New York: Macmillan, 1960), 106.
15. Stanley L. Jaki, "Medieval Christianity: Its Inventiveness in Technology and Science," in Arthur M. Melzer, Jerry Weinberger, and M. Richard Zinman, editors, *Technology in the Western Political Tradition* (New York: Cornell University Press, 1993), 67–68.

3

C. S. Lewis on Science as a Threat to Freedom

Edward J. Larson

Among other things, C. S. Lewis considered modern science a threat to freedom in modern society. In order to understand how modern science is a threat to freedom, we have to consider Lewis's view of modern science. Lewis lived at a time when science was emerging as the dominant system of thought in the Western world, when the technological spin-offs of that intellectual activity were fundamentally transforming every aspect of life. Lewis reflected on this in his 1954 inaugural lecture at Cambridge University, when he declared:

> The sciences long remained like a lion-cub whose gambols delighted its master in private; it had not yet tasted man's blood. All through the eighteenth century... science was not the business of Man because Man had not yet become the business of science. It dealt chiefly with the inanimate; and it threw off few technological byproducts. When Watt makes his engine, Darwin starts monkeying with the ancestry of Man, and Freud with his soul, then indeed the lion will have got out of its cage.[1]

As we know from the Narnia tales, a free lion does not pose a threat so long as it is true—like Aslan. But Lewis did not view science as a source of neutral truths about nature. For example, in *The Discarded Image*, Lewis wrote about the differences between the medieval and modern model of nature: "The most spectacular differences between the Medieval Model and our own concern astronomy and biology. In both fields the New Model is supported by a wealth of empirical evidence. But we should misrepresent the historical process if we said that the irruption of new facts was the sole cause of the alteration."[2] According to Lewis, "The old astronomy was not, in any exact sense, 'refuted' by the telescope… The new astronomy triumphed not because the cause for the old became desperate, but because the new was a better tool."[3] The same was true of the revolution in biology "from a devolutionary to an evolutionary scheme": "This revolution was certainly not brought about by the discovery of new facts… The demand for a developing world—a demand obviously in harmony both with the revolutionary and the romantic temper—grows up first; when it is full grown the scientists go to work and discover the evidence."[4] Lewis made similar comments about the development of Darwinian evolution in "The Funeral of a Great Myth," where he concluded, "every age gets, within certain limits, the science it desires."[5] Lewis noted that he did "not at all mean that these new phenomena are illusory," but he believed that "Nature has all sorts of phenomena in stock and can suit many different tastes."[6]

Viewing modern science as a reflection of its age, rather than a method for finding truth, does not necessarily transform it into a threat to freedom. But Lewis's dark foreboding about the direction of modern civilization inevitably cast a shadow over the sciences, which represent our civilization's defining achievement. At least three bases for this concern run through Lewis's writings.

1. THE TENDENCY TOWARD SCIENTIFIC REDUCTIONISM

FIRST, LEWIS FEARED THAT THE reductionist tendency of modern science undermined moral reasoning, human dignity, and religious faith. In a sweeping statement from his academic masterwork, *English Literature in the Sixteenth Century*, Lewis declared that modern science "substituted a mechanical for a genial or animistic conception of the universe. The world was emptied, first of her indwelling spirits, then of her occult sympathies and antipathies, finally of her colours, smells and tastes."[7] Focusing on the biological and social sciences in "The Funeral of a Great Myth," Lewis added that the modern theory of evolution "asks me to believe that reason is simply the unforeseen and unintended byproduct of a mindless process at one stage of its endless and aimless becoming."[8] Reason is thus viewed as a product of non-rational nature. This undermines moral reasoning because our moral judgments depend on our reasoning, and if our reasoning is not grounded in the rational, then neither are our moral judgments. Moral reasoning is stripped of any claims to be truth, in the rational sense. Accordingly, the scientific naturalist must say, as Lewis put it in *Miracles*, "there is no such thing as wrong and right, I admit that no moral judgment can be 'true' or 'correct' and, consequently, that no one system of morality can be better or worse than another."[9]

This line of thinking inevitably results in moral relativism, which diminishes human distinctiveness by asserting that human values, theories, and even religious beliefs are subjective rather than objective. As Lewis expressed it in *The Discarded Image*, "Always century by century, item after item is transferred from the object's side of the account to the subject's. And now, in some extreme forms of Behaviourism, the subject himself is discounted as merely subjective; we only think that we think. Having eaten up everything else, he eats himself up too. And where we 'go from that' is a dark question."[10]

Brooding on this dark question in "The Empty Universe," Lewis wrote of scientific reductionism, "While we were reducing the world to almost nothing we deceived ourselves with the fancy that all its lost qualities were being kept safe (if in a somewhat humbled condition) as 'things in our own mind'."[11] But he added, "just as we have been broken of our bad habit of personifying trees, so [modern science says] we must now be broken of our bad habit of personifying men... The Subject is as empty as the Object."[12] Religious faith is the ultimate victim of this way of thinking, as expressed in the words of Lewis's scientific protagonist in his essay "Religion and Science":

> 'Miracles,' said my friend, 'Oh, come. Science has knocked the bottom out of all that. We know that Nature is governed by fixed laws... I mean, the laws of Nature tell us not merely how things *do* happen but how they *must* happen. No power could possibly alter them... The whole picture of the universe which science has given us makes it such

rot to believe that the Power at the back of it all could be interested in us tiny little creatures crawling about on an unimportant planet!"[13]

2. The Challenges of Technology

SECOND, SCIENCE LEADS TO TECHNOLOGY, which Lewis believed would be utilized regardless of its detrimental impact on humans. In general, Lewis was neutral toward the so-called "advance" of modern technology; a neutrality that logically followed from his view, expressed in "The World's Last Night," that "[I]n my opinion, the modern conception of Progress... is simply a myth, supported by no evidence whatever."[14] So, in answer to the question, "Is progress even possible?" he wrote about technology: "We shall grow able to cure, and to produce, more diseases—bacterial war, not bombs, might ring down the curtain—to alleviate, and to inflict, more pains, to husband, or to waste, the resources of the planet more extensively. We can become neither more beneficent nor more mischievous. My guess is we shall do both."[15]

Lewis does acknowledge that, for good or for ill, technology gives humans more power over nature. In *That Hideous Strength*, he portrayed a brave new high-tech world where something ought to be done simply because it can be done. Thus, at one point, the science professor Augustus Frost explains that a particular activity "is justified by the fact that it is occurring, and ought to be increased because an increase is taking place." When Lewis's hero questions the moral implications of the activity, Frost replies, "The judgment you are trying to make turns out on inspection to be simply an expression of emotion."[16] So viewed, technology is not answerable to any higher standard. Its sole parameter is the possible.

3. Science in Service of Oppression

THIRD, LEWIS WAS CONVINCED THAT scientific authority would be used to justify and facilitate political oppression. Also in *That Hideous Strength*, Lewis observed, "The physical sciences, good and innocent in themselves, had already... begun to be warped, and been subtly manoeuvered in a certain direction. Despair of objective truth had been increasingly insinuated into the scientists; indifference to it, and a concentration upon mere power, had been the result."[17] For Lewis, the threat here is quite real. Commenting on this book, which is so damning of modern science, Lewis later wrote, "'scientists' as such are not the target... what we are obviously up against throughout the story is not scientists but officials."[18] This is an important distinction for Lewis. Scientific planning is not necessarily evil, "but 'Under modern conditions any effective invitation to Hell will certainly appear in the guise of scientific planning'—as Hitler's regime in fact did."[19]

Elaborating on this theme in his essay "Is Progress Possible?," Lewis concluded that "the new oligarchy must more and more base its claim to plan us on its claim of knowledge... This means they must increasingly rely on the advice of scientists."[20] Lewis added that he "dread[ed] specialists in power because they are specialists speaking outside their special subjects. Let scientists tell us about science. But government involves questions about the good of man, and justice, and what things are worth having at what price; and on these a scientific training gives a man's opinion no added value." This is why Lewis feared "government in the name of science. That is how tyrannies come in. In every age the men who want us under their thumb, if they have any sense, will put forward the particular pretension which the hopes and fears of that age render most potent... It has been magic, it has been Christianity. Now it will certainly be science."[21]

In this sense, Lewis perceived science as the ultimate threat to freedom in modern society.

Endnotes

1. C. S. Lewis, "*De Descriptione Temporum*," in Lyle W. Dorsett, editor, *The Essential C. S. Lewis* (New York: Touchstone, 1996), 476.
2. C. S. Lewis, *The Discarded Image* (Cambridge: Cambridge University Press, 1964), 219.
3. Ibid., 219–220.
4. Ibid., 220–221.

5. C. S. Lewis, "The Funeral of a Great Myth," in *Christian Reflections*, edited by Walter Hooper (Grand Rapids: Eerdmans, 1967), 85.
6. Lewis, *The Discarded Image*, 221.
7. C. S. Lewis, *English History in the Sixteenth Century Excluding Drama* (London: Oxford University Press, 1954), 3.
8. Lewis, "Funeral of a Great Myth," 89.
9. C. S. Lewis, *Miracles: A Preliminary Study*, revised edition (New York: Macmillan, 1978), 36.
10. Lewis, *The Discarded Image*, 215.
11. C. S. Lewis, "The Empty Universe," in *Present Concerns*, edited by Walter Hooper (San Diego: HBJ, 1986), 83.
12. Ibid., 82–83.
13. C. S. Lewis, "Religion and Science," in *God in the Dock: Essays on Theology and Ethics*, edited by Walter Hooper (Grand Rapids: Eerdmans, 1970), 72–74.
14. C. S. Lewis, "The World's Last Night," in C. S. Lewis, *The World's Last Night and Other Essays* (San Diego: Harcourt Brace and Co., 1960), 101.
15. C. S. Lewis, "Is Progress Possible?" in *God in the Dock*, 312.
16. C. S. Lewis, *That Hideous Strength* (New York: Macmillan, 1965), 295.
17. Ibid., 203.
18. C. S. Lewis, "A Reply to Professor Haldane," in C. S. Lewis, *Of Other Worlds: Essays and Stories*, edited by Walter Hooper (New York: Harcourt Brace Jovanovich, 1966), 78.
19. Ibid., 80.
20. C. S. Lewis, "Is Progress Possible?" in *God in the Dock*, 314.
21. Ibid., 315.

4

C. S. Lewis, Science, and the Medieval Mind

Jake Akins

The persecution of Galileo. The belief the Earth is flat. The burning of witches. During the past century, these iconic examples have been pressed into service time and again to dismiss the Middle Ages as an era of anti-scientific nonsense. This gloomy depiction of medieval culture drew from a much larger cultural narrative, popular since the nineteenth century, which presented belief in the supernatural as an obstacle to the steady march of scientific progress.

One of C. S. Lewis's signal contributions to the debate over science in modern life was to disabuse people of this sort of "chronological snobbery"[1] toward the Middle Ages and its supposedly anti-scientific view of the natural world. But ultimately Lewis did much more than merely debunk errors about the Middle Ages. He also showed the positive contribution made by medieval thinkers to modern scientific methodology, and he revealed how the revolution from medieval to modern science exposed human frailties inherent in the scientific enterprise that need to be understood if modern science is to continue to flourish.

Everything You Thought You Knew about the Middle Ages Is Wrong

Of all the ideas in the popular imagination that purportedly prove the Middle Ages was "anti-science," the claim that medieval thinkers believed the Earth was flat is surely one of the most hardy and perennial. "The erroneous notion that the medievals were Flat-earthers was common enough until recently," wrote Lewis in his last book, *The Discarded*

Image.[2] In reality, he explained, "all the authors of the high Middle Ages" agreed that "the Earth is a globe." Moreover, "the implications of a spherical Earth were fully grasped. What we call gravitation… was a matter of common knowledge."[3] Lewis acknowledged that there were a handful of fringe figures who may not have embraced a spherical Earth, but he noted that one could find such outliers even "in the nineteenth century."[4] The fact remains that medieval culture as a whole embraced a spherical Earth, as did earlier Greco-Roman thinkers such as Cicero.

A second myth used to tar medieval thinkers as unscientific is the claim that they wrongly "thought the Earth was the largest thing" in the universe, which misled them into thinking that "the Creator was specially interested in Man and might even interrupt the course of Nature for his benefit." Lewis remarked that "[w]hatever its value may be as an argument… this view is quite wrong about facts." He went on to point out that "[m]ore than seventeen hundred years ago Ptolemy taught that in relation to the distance of the fixed stars the whole Earth must be regarded as a point with no magnitude. His astronomical system was universally accepted in the Dark and Middle Ages." Thus, "[t]he insignificance of Earth was as much a commonplace to Boethius, King Alfred, Dante, and Chaucer as it is to Mr. H. G. Wells or Professor Haldane."[5]

A third slur on the medievals' understanding of the natural world is the claim that medieval science personified material objects by crudely attributing to them willful behavior. In *The Discarded Image*, Lewis points out that the "fundamental concept of modern science is, or was till very recently, that of natural 'laws', and every event was described as happening in 'obedience' to them." In medieval science, by contrast, "the fundamental concept was that of certain sympathies, antipathies, and strivings inherent in matter itself. Everything has its right place, its home, the region that suits it, and, if not forcibly restrained, moves thither by a sort of homing instinct."[6] In Lewis's book *Studies in Medieval and Renaissance Literature* we find the same characterization: "Their way of describing it is to say that every natural object has a native or 'proper' place and is always 'trying' or 'desiring' to get there. When unimpeded, flame moves upwards and solid bodies move downwards because they want to go, you may call it 'home.'"[7]

Anticipating the modern response to this conception, Lewis asks: "Did they really think all matter was sentient? Apparently not. They will distinguish animate and inanimate as clearly as we do; will say that stones, for example, have only being; vegetables being and life; animals, being, life and sense; man, being, life, sense and reason."[8]

Lewis makes the point that the medievals used this language as an analogy in the same way that moderns use the language of objects "obeying the laws of nature" as an analogy. Lewis points out that if a modern man were to ask a medievalist if he really believed inanimate objects to "desire home," the medievalist could respond by asking whether modern men really believed that inanimate objects could "obey laws." Both ways of speaking personify inanimate objects, and modern terminology is arguably more blatant in doing so according to Lewis: "The odd thing is that ours is the more anthropomorphic of the two. To talk as if inanimate bodies had a homing instinct is to bring them no nearer to us than the pigeons; to talk as if they could 'obey laws' is to treat them like men and even like citizens."[9]

Magic, Astrology, and the Middle Ages

But what about the flourishing of magic and astrology during the Middle Ages? Surely those endeavors indicated a culture in the thrall of superstition and blatantly anti-scientific? Not according to Lewis. "There was very little magic in the Middle Ages," he insisted, arguing that it was the centuries following the Middle Ages that constituted "the high noon of magic."[10]

Lewis did acknowledge the flourishing of astrology during the Middle Ages, but he contended that this was not a very good example of magical, let alone supernatural, thinking about nature. To understand why, we need to go into more detail.[11]

In the Middle Ages, planets were (wrongly) thought to orbit Earth. The medievals further believed that planets had influence over the course of events on Earth and influence over human psychology. The planets influenced Earth through the æther believed to be above the Moon and through the air between the Moon and Earth. Certain positions of the planets were believed to make the air healthy or unhealthy. The Italian word for influence is *influenza*. States of the planets and their influence on Earth were believed to cause people to become ill in the unhealthy air, and thus get influenza, commonly referred to as the flu. Medievals believed that the planets' beams permeated the Earth's crust, and finding suitable soil, produced certain metals (e.g. Saturn produced lead, Jove tin, the Sun gold, etc.). Jove is, of course, Jupiter. The medieval belief in planetary beams changing soil to metal is related to the practice of alchemy, a discipline that continued to be pursued well after the Middle Ages by such noted figures as Sir Isaac Newton, considered one of the greatest scientists of all time.

According to the medievals, this influence of the planets did not in the end compel humans to think certain thoughts. The planets merely produced predispositions, such as being melancholy if one was born under Saturn, or amorous under Venus. The phases of the moon were thought to have a negative influence on man's normal sense of reason, which is the derivation of our word "lunacy" and "lunatic." It was believed that one could overcome this planetary influence as one can overcome a bad temper. Nevertheless, while this influence did not negate individual free will, it did exercise such an influence on a person's state of mind that astrologers were believed able to make predictions based on astrological considerations.

What is intriguing is that these predictions of astrology were not thought to be "magic." Indeed, according to Lewis, this is the crucial difference between modern astrology and medieval astrology. Modern astrology is associated with magic, whereas medieval astrology was the practice of discerning how heavenly bodies influenced our world, such as how the Moon influences the ocean's tide. Medieval astrology was thought of not as magic but as a hard science, as physics or chemistry is thought of today.

As a result, "[a]strology was a hardheaded, stern, anti-idealistic affair; the creed of men who wanted a universe which admitted no incalculables." Unlike magic, in which men "sought power over nature," "astrology proclaimed nature's power over man." That is why in Lewis's view the medieval astrologer is not appropriately connected with the later interest in magic. Nor is the astrologer connected to the supernatural thinking of Christianity, which, if anything, undermined claims of "astrological determinism" as well as predictions of the future based on astrology. Instead, the medieval astrologer according to Lewis is more properly regarded as the ancestor of "the nineteenth century philosophical materialist."[12] Just as the scientific materialists of the 1800s claimed to reduce all human behavior to the product of a blind deterministic process grounded in nature, so too did the medieval astrologers. As a result, it might even be said that medieval astrology helped pave the way for the materialistic reductionism in science that became so prevalent in the centuries following the Middle Ages.

Lewis viewed the contribution of the Middle Ages to the rise of modern scientific determinism with trepidation, as he made clear in works like *The Abolition of Man* and *That Hideous Strength*. But there was a much more positive contribution made by medieval thinkers to the development of the scientific way of thinking that Lewis thought had continuing relevance for understanding the limits of science today.

The Medieval Origin of Scientific Methodology

Lewis illuminates two elements of science implemented in the Middle Ages that will put into perspective the advanced scientific sophistication of the medievals—"saving the appearances" and "Ockham's Razor." Lewis explains the concept of "saving the appearances":

> A scientific theory must 'save' or 'preserve' the appearances, the phenomena, it deals with, in the sense of getting them all in, doing justice to them. Thus, for example, your phenomena are luminous points in the night sky which exhibit such and such movements in relation to one another and in relation to an observer at a particular point, or various chosen points, on the surface of the Earth. Your astronomical theory

will be a supposal such that, if it were true, the apparent motions from the point or points of observation would be those you have actually observed. The theory will then have 'got in' or 'saved' the appearances.[13]

This idea of "saving the appearances" in the origin of science influenced Lewis after he read his friend Owen Barfield's book *Saving the Appearances: A Study in Idolatry*. The concept holds that all of the relevant phenomena need to be explained by a scientific hypothesis, or else the hypothesis isn't doing justice to the total phenomena; and the hypothesis cannot simply explain away the phenomena, or else the phenomena will not have been "saved."

This is a truism in science in general. However, Lewis pointed out that more needs to be added to this criterion in order to provide a satisfactory scientific theory, for "if we demanded no more than that from a theory, science would be impossible, for a lively inventive faculty could devise a good many different supposals which would equally save the phenomena." The additional criterion according to Lewis was "first, perhaps, formulated with full clarity by" philosopher William of Ockham (c. 1287–1347). "According to this second canon we must accept (provisionally) not any theory which saves the phenomena but that theory which does so with the fewest possible assumptions."[14] Today this idea is commonly known as "Ockham's Razor."

Given these two criteria, Lewis pointed out that scientists didn't then, and modern scientists shouldn't now, consider scientific theories to be statements of fact. The particulars being studied are facts, but the theory is only provisional, i.e., "it will have to be abandoned if a more ingenious person thinks of a supposal which would 'save' the observed phenomena with still fewer assumptions, or if we discover new phenomena which it cannot save at all."[15]

An interesting instance of this criterion being neglected in the history of science which modern readers will be familiar with is what actually caused the uproar during the Copernican Revolution. According to Lewis, "The real reason why Copernicus raised no ripple and Galileo raised a storm, may well be that whereas the one offered a new supposal about celestial motions, the other insisted on treating the supposal as fact. If so, the real revolution consisted not in a new theory of the heavens but in 'a new theory of the nature of theory'."[16] Here Lewis is referencing Barfield's *Saving the Appearances*, which states:

> The real turning point in the history of astronomy and of science in general was something else altogether. It took place when Copernicus (probably—it cannot be regarded as certain) began to think, and others, like Kepler and Galileo, began to affirm that the heliocentric hypothesis not only saved the appearances, but was physically true. It was this, this novel idea that the Copernican (and therefore any other) hypothesis might not be a hypothesis at all but the ultimate truth, that was almost enough to constitute the 'scientific revolution'... It was not simply a new theory of the nature of celestial movements that was feared, but a new theory of the nature of theory; namely, that, if a hypothesis saves all the appearances, it is identical with truth.[17]

Contrary to the modern belief that the Roman Catholic Church resisted heliocentrism on religious grounds, the real story is that the Church realized that the implication in treating a hypothesis as fact unjustifiably leaves out the possibility of a better hypothesis that saves the appearances and has even fewer assumptions. The Church's position actually allowed for further scientific investigation. Indeed, according to Barfield, it was the Church that encouraged Copernicus to publish his work that proposed the heliocentric model. Copernicus had already been teaching the heliocentric hypothesis with no issue with the Church, because he treated it as a hypothesis. In this way, the medievals were more thoughtful in regarding scientific hypotheses as provisional than the modern popular conception, which tends to believe that modern scientific declarations are identical with truth.

THE PSYCHOLOGY OF SCIENTIFIC REVOLUTIONS

THE MEDIEVAL MIND'S RECOGNITION OF the provisional nature of scientific knowledge is perhaps its most important lesson for us today.

Lewis makes this lesson clear in his epilogue to *The Discarded Image*, where he acknowledges that although the medieval model of the universe "delights" him, "it had a serious defect; it was not true."[18] He goes on to argue that the medieval model as a whole didn't fold under the weight of growing and refined scientific evidence, though a good many particulars were replaced (just as the modern model's particulars are being replaced today in the laboratory), but rather the model was discarded due to changes in man's psychology.

This was true in the scientific revolution that replaced medieval astronomy, and it was especially true in the scientific revolution that overthrew medieval biology. According to Lewis, biology's metamorphosis "from a devolutionary to an evolutionary scheme; from a cosmology in which it was axiomatic that 'all perfect things precede all imperfect things' to one in which it is axiomatic that 'the starting point (*Entwicklungsgrund*) is always lower than what is developed'"[19] did not derive primarily from new scientific evidence but from a change in cultural attitudes: "The demand for a developing world—a demand obviously in harmony both with the revolutionary and the romantic temper—grows up first; when it is full grown the scientists go to work and discover the evidence." Thus, "[t]here is no question here of the old Model's being shattered by the inrush of new phenomena. The truth would seem to be the reverse; that when changes in the human mind produce a sufficient disrelish of the old Model and a sufficient hankering for some new one, phenomena to support that new one will obediently turn up."[20]

As a consequence, Lewis believed that any model, when taken as a whole, should only be held provisionally, with the understanding that our philosophical presuppositions and worldviews play a large part in formulating the model itself, and with the understanding that as an age's philosophies and worldviews change so will its model of nature.

Lewis did not advocate a return to the medieval model, but he was "suggesting considerations that may induce us to regard all Models in the right way, respecting each and idolizing none," for "[w]e are all, very properly, familiar with the idea that in every age the human mind is deeply influenced by the accepted Model of the universe. But there is a two-way traffic; the Model is also influenced by the prevailing temper of mind."[21] Therefore, "[w]e can no longer dismiss the change of Models as a simple progress from error to truth. No Model is a catalogue of ultimate realities, and none is a mere fantasy. Each is a serious attempt to get in all the phenomena known at a given period... But also, no less surely, each reflects the prevalent psychology of an age almost as much as it reflects the state of that age's knowledge."[22]

Lewis believed that our own model of the universe will be replaced eventually by a new one when our descendants' psychology changes to a large enough degree. "The new Model," he explained, "will not be set up without evidence, but the evidence will turn up when the inner need for it becomes sufficiently great."[23]

Lewis's view of the psychology of scientific revolutions might be regarded by some as an attack of the validity of modern science. But it is more accurately described as a rejection of scientific dogmatism and a plea for continuing scientific exploration. Rather than impeding the work of scientists, Lewis's view invites scientists to question their current assumptions and to think outside the box. It nurtures scientific humility. It encourages a healthy tolerance of those who may dissent from the reigning scientific orthodoxy. These are qualities that are sorely needed in many current debates over science in the public square.

Endnotes

1. C. S. Lewis, *Surprised by Joy* in *The Inspirational Writings of C. S. Lewis* (Orlando, Florida: Harcourt Brace & Company, 1986), 114.
2. C. S. Lewis, *The Discarded Image, An Introduction to Medieval and Renaissance Literature* (United Kingdom: Cambridge University Press, 2007), 142.
3. Ibid., 140, 141.
4. Ibid., 140.
5. Ibid., 97; see also C. S. Lewis, *Studies in Medieval and Renaissance Literature*, collected by Walter Hooper (Cambridge: Cambridge University Press, 2007), 46.
6. Lewis, *The Discarded Image*, 92–93.
7. Lewis, *Studies in Medieval and Renaissance Literature*, 49.
8. Ibid., 49–50.
9. Lewis, *The Discarded Image*, 94.
10. C. S. Lewis, *The Abolition of Man* (New York: Macmillan, 1955), 87.

11. The following discussion draws on and summarizes the views Lewis presented in *Studies in Medieval and Renaissance Literature*, 54–57.
12. Ibid., 55–56.
13. Lewis, *The Discarded Image*, 14–15.
14. Ibid., 15.
15. Ibid., 16.
16. Ibid., 16.
17. Owen Barfield, *Saving the Appearances: A Study in Idolatry* (Middletown, Connecticut: Wesleyan University Press, 1988), 50–51.
18. Lewis, *The Discarded Image*, 216.
19. Ibid., 220.
20. Ibid., 221.
21. Ibid., 222.
22. Ibid., 222.
23. Ibid., 223.

5

A Peculiar Clarity

How C. S. Lewis Can Help Us Think about Faith and Science

C. John Collins

Famous both as a defender of Christian faith and as the writer of imaginative fiction, C. S. Lewis actually had a day job: He was a professional scholar of medieval and Renaissance European literature. From 1925 until 1954, he was Fellow and Tutor in English Literature at Magdalen College, Oxford; and from 1954 until he retired in 1963 (shortly before he died that same year), he was Professor of Medieval and Renaissance Literature at Cambridge University. In the course of his academic work he produced books and papers on topics in ideological history, English philology, and literary interpretation, many of which still show considerable value.

Even though the general public knows Lewis primarily for the theological, apologetic, and imaginative works, nevertheless his total work hangs together; the same personal traits come through in all of his writings.

At one point Lewis did consider becoming an academic philosopher: During 1924–25, he held a temporary position as tutor in philosophy on behalf of E. F. Carritt (philosophy tutor at Oxford's University College), who spent the year teaching at the University of Michigan. However, he could not endure that as a career; he wrote to his father: "A continued search among the abstract roots of things, a perpetual questioning of all that plain men take for granted, a chewing the cud for fifty years over inevitable ignorance and a constant frontier watch on the little tidy conventional world of science and daily life—is this the best life for temperaments such as ours?"[1]

Perhaps surprisingly, many philosophers have found in Lewis worthy material for their serious discussions.[2] By way of parallel, Lewis was not a professional theologian or Biblical scholar, but exegetes and theologians have still found themselves stimulated by what he wrote.[3]

Lewis the apologist wrote during a time when the standard narrative in the Western world was that the advances of the sciences were relegating the archaic beliefs of traditional religions such as Christianity to the museum.[4] As Christians sought to adapt to the new knowledge, many skeptics held these efforts in contempt.[5] Lewis took it to be his job to defend the essentials of Christian belief, and to show that these essentials wear well as they encounter modern trends of thought.

Lewis appeals to all manner of Christians—Protestant, Roman Catholic, and Eastern Orthodox. Roger Lancelyn Green and Walter Hooper record an amusing incident from 1954 in which the American fundamentalist Bob Jones, Jr, expressed his (grudging) approval of Lewis.[6] Even some people, whom we might call "broadly theist"—who accept that there is a God, but are not Christians—have spoken appreciatively of Lewis.[7] More recently, the prominent Christian geneticist Francis Collins has attributed much of his own spiritual journey toward Christian faith to Lewis's influence, weaving references to Lewis's apologetics all through his 2006 testimony, *The Language of God*.[8] Collins articulates a version of "theistic evolution" that he calls "BioLogos," and helped to establish the BioLogos Foundation. Visitors to the foundation's web site (biologos.org) will find essays that invoke Lewis as an ally in supporting this perspective.[9]

I have been thinking and writing, for a number of years, about how science and faith interact. Lewis has supplied me with helpful ideas more than anyone else.[10] I began reading his works during January of 1974, when I was a second-year student in computer science and electrical engineering at MIT; his apologetics, his fantasy series, and his science fiction all appealed to me. Later, when I was doing doctoral research in issues of methodology in comparative Semitic lexicography (that is, in considering how the meanings of related words in different Semitic languages, such as Hebrew and Arabic, might help us read ancient texts in these languages),[11] I found that Lewis's *Studies in Words* provided an excellent example of good lexical method applied to reading old texts without getting bogged down in theoretical fine points.[12] I do not claim that I agree with every argument or formulation that Lewis offered; rather, I find that what we may call a "broadly Lewisian" outlook continues to provide the intellectual tools for my thinking.[13]

What is it about Lewis's approach that makes him, a non-specialist, so helpful, and particularly helpful in thinking about how science and faith relate to one another? I consider the key to be his knack for clarity of thought. My purpose in this essay is therefore to explore some aspects of Lewis's background and style of thought that provide this clarity, aspects that we might strive to emulate.[14]

1. Historical Depth

THE PERSON WHO READS WHAT Lewis wrote as a professional specialist will discover that his Oxford education had given him a clear and detailed grasp of the history of European thought, a thought with roots both in the Greco-Roman classics and in Judeo-Christian faith.

An obvious value of Lewis's grasp of this history is his ability to recognize and expose instances of misinformation, especially when that misinformation was being used for propaganda purposes. For example, the pre-eminent evolutionary biologist (and anti-Christian) J. B. S. Haldane had written in his popular work, *Possible Worlds* (1927): "Five hundred years ago the human mind was limited to a tiny patch of space, and the universe must have seemed even smaller after Magellan's men had girdled the earth. The heavenly bodies were known to be distant, but it was not clear that celestial distances were so much greater than terrestrial."[15]

Lewis's *Discarded Image*, published posthumously in 1964, was based on his university lectures introducing the world picture assumed by medieval literature. He observed there that "Casual statements about pre-Copernican astronomy in modern scientists who are not historians are often unreliable."[16] Later in the book he identifies Haldane outright as a chief source of such "unreliable" statements, and he explains how medieval thinkers knew that the Earth in cosmic terms "had no appreciable magnitude" and that the distances between the Earth and the stars were (as we might put it today) astronomical.[17] Lewis must have thought that Haldane's assertion about the medieval world picture was both influential and dangerous, because he mentions the assertion and its refutation several times in his writings and talks.[18]

Depth like Lewis's can protect us from incorrect historical assertions, which are designed to lead us to adopt or abandon some line of thinking. For example, most people are aware that in medieval cosmology, the earth was taken to be stationary, with the rest of the observable universe moving around it. But it is exceedingly common to find authors mistaking this physical location for something more ideological; that is, we read that the researches of Copernicus and Galileo displaced the earth, and thus humankind, from their central position in God's plan. An article in the *Christian Science Monitor* on the 374th anniversary of the birth of Nicholas Steno (a devout Christian and in some ways the father of modern geology) took just this tack: "Just as the findings of Copernicus and the astronomers that followed him revealed that the earth is not the hub of the universe, Steno's revolution dislodged humanity from the center of our planet's history."[19]

Jay Richards contended in reply:

> This trope functions like a keyboard macro when journalists write about the history of science. But it's pure mythology. In pre-Copernican cosmology, the earth was *not* seen as the hub of the universe, but as the bottom, the place to which heavy, mutable things fall. The very center of the universe, it was supposed, was Hell—hardly a place of privilege. This idea that the center must be the place of privilege is modernist misinterpretation of the history of science. Copernicus most assuredly did *not* "dislodge humanity" from the imagined hub of the universe. And neither, of course, did Steno. How exactly does the moment we

arrive on the scene settle questions about our importance? If a bride arrives at her wedding in the last hour, even though preparations have been underway for a year, does that mean she's insignificant?[20]

A perusal of Lewis's *Discarded Image* (particularly chapter 5) shows that Richards has the better grasp of the *meaning* of the Earth's location in the medieval view.

But there was certainly more to Lewis's historical depth than simply accurate knowledge of what people thought. His historical perspective on the changes in European thought led him to some conclusions about how people come to replace grand scientific theories with newer ones. He recognized a subtle interplay between new data, worldview, and human nature, putting it all in a way that Thomas Kuhn might well have wished he had used.[21] As Lewis observed:

> The old astronomy was not, in any exact sense, 'refuted' by the telescope. The scarred surface of the Moon and the satellites of Jupiter can, if one wants, be fitted into a geocentric scheme... How far, by endless tinkerings, it could have kept up with them till even now, I do not know. But the human mind will not long endure such ever-increasing complications if once it has seen that some simpler conception can 'save the appearances'... The new astronomy triumphed not because the case for the old became desperate, but because the new was a better tool; once this was grasped, our ingrained conviction that Nature herself is thrifty did the rest.[22]

Some might wonder whether Lewis was an anti-realist when it comes to science (the same question arises with Kuhn). I suspect that the answer is not really (see point 2 below); rather, he distinguished between the more modest results on the smaller scale (dealing with observations, experiments, measurements, and well-grounded inferences), and grand-scale unifying theories (which he called Models). As he wrote:

> No Model is a catalogue of ultimate realities, and none is a mere fantasy. Each is a serious attempt to get in all the phenomena known at a given period, and each succeeds in getting in a great many. But also, no less surely, each reflects the prevalent psychology of an age almost as much as it reflects the state of that age's knowledge.[23]

Lewis's historical sense enabled him to recognize what he came to call "chronological snobbery": the "uncritical acceptance of the intellectual climate common to our own age and the assumption that whatever has gone out of date is on that account discredited."[24] Understanding the danger of "chronological snobbery" allows us both to admire the achievements (including scientific ones) of earlier ages, and to exercise proper critique of our own pretensions to superior cleverness.

Besides providing the sympathy needed for good historiography, this stance enables us to read the ancient texts on their own terms.[25] For example, when Lewis compares the medieval view that there are "certain sympathies, antipathies, and strivings inherent in matter itself," with the modern scientific view that events "obey" natural "laws," he does not simply dismiss the medieval view as outdated.[26] Rather, he points out the metaphorical underpinnings of both views:

> The question at once arises whether medieval thinkers really believed that what we now call inanimate objects were sentient and purposive. The answer in general is undoubtedly no… If we could ask the medieval scientist 'Why, then, do you talk as if they did', he might (for he was always a dialectician) retort with the counter-question, 'But do you intend your language about *laws* and *obedience* any more literally than I intend mine about *kindly enclyning?*'[27]

Lewis's historical depth ultimately enabled him to appreciate how "science" came to be the chief preoccupation of educated Western people. Prior to the nineteenth century, he pointed out, "[s]cience was not the business of Man because Man had not yet become the business of science. It dealt chiefly with the inanimate; and it threw off few technological by-products." But "[w]hen Watt makes his engine, when Darwin starts monkeying with the ancestry of Man, and Freud with his soul, and the economists with all that is his, then indeed the lion will have got out of its cage."[28]

2. Respect for Both the Benefits and the Limits of Science

The Use and Abuse of Science

The early twentieth century represents a time in English-speaking culture in which the sciences were coming more and more into prominence, and beginning to drive technological developments, with manifest improvements in the lives of many people. In such a setting, the question naturally arises, "Why don't we use 'scientific' approaches to improving *human nature* itself?" Such a prospect alarmed Lewis, because it leaves out of consideration any ethical reflections on what makes human life distinctive and valuable, largely because (so Lewis thought) it proceeded under a supposedly "scientific" (but really materialist) notion of humanity.

Lewis's argument in *The Abolition of Man* is directed against this plan for the sciences.[29] His basic thesis is "the doctrine of objective value," that is, "the belief that certain attitudes are really true, and others really false, to the kind of thing the universe is and the kind of things we are."[30] This system of values he calls the *Tao*. Lewis observes that "'Man's conquest of Nature' is an expression often used to describe the progress of applied science."[31] But unless science, like other human activities, is governed by a sound sense of what it means to be human, it can be employed to eliminate "human nature."

Lewis is clear that he does not oppose science as such.[32] Indeed, he even suggests "that from Science herself the cure might come."[33] Science works best when it is used to explain things, without aiming to explain them away. In other words, science cannot be set up on its own as the supreme arbiter of truth and value. Lewis makes clear that science, to be worthwhile, depends on the validity of reasoning—and this makes science subject to the requirements of sound critical thinking.[34] Further, science does not displace ethical reasoning: "Let scientists tell us about sciences. But government involves questions about the good for man and justice, and what things are worth having at what price; and on these a scientific training gives a man's opinion no added value."[35]

The upshot of this is that when someone wants to declare that a truly "scientific" approach must assume naturalism, or materialism, or reductionism—whether in the strong form of metaphysical assertion or in the weaker form of methodological prescription—that declaration is subject to review for its reasonableness. There may well be areas of thought in which the assumption is a fair one, mind you; but we must validate the assumption.

Indeed, Lewis was persuaded that the sciences will flourish in a theistic intellectual environment; in fact, he went so far as to contend that the metaphysical positions of naturalism and materialism actually undercut our ability to justify our scientific practice.[36] In particular, the validity of human reasoning depends on our acts of reasoning not being "interlocked with the total interlocking system of Nature as all its other items are interlocked with one another."[37] This is his famous "argument from reason," and we can sum it up as he did, with a quotation from a source that is usually hostile to Lewis, J. B. S. Haldane: "If my mental processes are determined wholly by the motions of atoms in my brain, I have no reason to suppose that my beliefs are true… and hence I have no reason for supposing my brain to be composed of atoms."[38]

To be confident that our reasoning can actually bring us in touch with *truth*—whether it be about science, or about moral questions, or anything else—requires that there be more than matter, acting according to its properties, in our thought processes.[39] Now this something more, the *super*-natural, is not the same as Christianity, though it leads at least to theism. Theism, which recognizes a supreme God who is the source of all being, who made humankind with capacities that go beyond the merely natural outgrowth of their material components—capacities that enable humans to understand and manage (even if only partially) the other things in the world—is the proper soil for science to grow in, according to Lewis:

5 A Peculiar Clarity / 77

If Naturalism is true we have no reason to trust our conviction that Nature is uniform. It can be trusted only if quite a different metaphysic is true… The sciences logically require a metaphysic of this sort. Our greatest natural philosopher [Alfred North Whitehead] thinks it is also the metaphysic out of which they originally grew.

…

But if we admit God, must we admit Miracle? Indeed, indeed, you have no security against it. That is the bargain. Theology says to you in effect, "Admit God and with Him the risk of a few miracles, and I in return will ratify your faith in uniformity as regards the overwhelming majority of events."[40]

Lewis showed that it is Christian orthodoxy, including its affirmation of miracles, that grounds a positive view of the scientific endeavor.

Lewis made a helpful distinction between the results of the particular sciences and the ideological extrapolations, or "myths," people might want to draw from those results. In his day it was popular to argue that the science of biological evolution leads to a general outlook that he called "evolutionism," the notion that the world-process is continually developing—with the associated admonition that ethical behavior is that which gets with this developmental process. Lewis insisted:

I do not mean that the doctrine of Evolution as held by practising biologists is a Myth. It may be shown, by later biologists, to be a less satisfactory hypothesis than was hoped fifty years ago. But that does not amount to being a Myth. It is a genuine scientific hypothesis. But we must sharply distinguish between Evolution as a biological theorem and popular Evolutionism or Developmentalism which is certainly a Myth.[41]

In the same way, Lewis was realistic about how Christians should use the sciences in defending their faith. He was aware of the temptation we all face owing to the prestige of science: "Each new [scientific] discovery, even every new theory, is held at first to have the most wide-reaching theological and philosophical consequences. It is seized on by unbelievers as the basis for a new attack on Christianity; it is often, and more embarrassingly, seized by injudicious believers as the basis for a new defence." But Lewis added that "usually... when the popular hubbub has subsided and the novelty has been chewed over by real theologians, real scientists and real philosophers, both sides find themselves pretty much where they were before."[42] He accordingly advised those who want to defend the Christian faith to avoid "Sentences beginning 'Science has now proved'" because "If we try to base our apologetic on some recent development in science, we shall usually find that just as we have put the finishing touches to our argument science has changed its mind and quietly withdrawn the theory we have been using as our foundation stone."[43]

At the same time, Lewis urged Christians to keep up with scientific developments because Christians "have to answer the current scientific attitude towards Christianity, not the attitude which sciences adopted one hundred years ago." He even encouraged Christians to write books about science where the Christianity would be "latent, not explicit" in order to steer people away from materialism.[44] And he was not averse to making judicious use of current scientific developments himself. In Lewis's day, the kinds of cosmological theories we now call Big Bang theories were gaining traction, and he was willing to point out the theistic implications:

> If anything emerges clearly from modern physics, it is that nature is not everlasting. The universe had a beginning, and will have an end. But the great materialistic systems of the past all believed in the eternity, and thence in the self-existence of matter... This fundamental ground for materialism has now been withdrawn. We should not lean too heavily on this, for scientific theories change. But at the moment it appears that the burden of proof rests, not on us, but on those who deny that nature has some cause beyond herself.[45]

The Dangers of Scientific Utopianism

WHEN LEWIS wrote, it was becoming common to trust scientific advancement to enable humans to travel in space, and even to colonize other worlds in the pursuit of a scientific utopia. Some advocates of this perspective tied it to their notion of what Lewis sometimes called Evolutionism, sometimes "scientism."[46] This forms the backdrop for Lewis's space trilogy, especially the first story (*Out of the Silent Planet*, 1938) in the person of the physicist Professor Weston. Elsewhere Lewis describes another version of this futuristic expectation: "A race of demigods now rules the planet—and perhaps more than the planet—for eugenics have made certain that only demigods will be born, and psychoanalysis that none of them shall lose or besmirch his divinity, and communism that all which divinity requires shall be ready to their hands."[47]

Chief among the popularizers of such views were H. G. Wells and J. B. S. Haldane. Lewis even preferred calling this approach "Wellsianity" rather than the "Scientific Outlook" (since it defamed honest science).[48] Not surprisingly, Haldane wrote a scathing review of the *Space Trilogy*, entitled "Auld Hornie, FRS" (1946).[49] Lewis drafted a reply, which was not published in his lifetime. There he observes that "[t]he odd thing is that Professor Haldane thinks Weston 'recognisable as a scientist'. I am relieved, for I had my doubts about him."[50]

Rejecting Haldane's claim that he was opposed to science, Lewis described a character in the third book of the trilogy, *That Hideous Strength*: "The 'good' scientist is put in precisely to show that 'scientists' as such are not the target."[51] Lewis held that science was being misused, owing to its prestige in European culture. He even summed up the "message" behind his third book: "But if you must reduce the romance to a proposition, the proposition would be… 'Under modern conditions any effective invitation to Hell will certainly appear in the guise of scientific planning.'"[52]

That Lewis was on target, and that people ignore his warning at their peril, is obvious from the history of the eugenics movement in the United States, where people were assessed for intelligence, and sometimes sterilized, all in the name of good science (e.g., "intelligence" tests, protecting the gene pool).[53]

Distinguishing Science from Non-Science

IN RECENT years, philosophers of science have realized that one of the difficulties in talking about "science" lies in defining just what is and is not "science"; this has been called the "demarcation problem." Lewis wrote at a time when a fairly simplistic account of scientists and their work was acceptable, and it shows. For example: "The scientist studies those elements in reality which repeat themselves. The historian studies the unique."[54]

Here Lewis is describing the subset of science that some call "nomothetic," because it is concerned with "laws" (Greek *nomoi*) or regularities. This leaves no room for those sciences that have an historical component, such as cosmology, geology, and evolutionary biology—all of which Lewis counted as sciences without any apparent hesitation, and all of which can deal with unique events.[55] Indeed, Lewis explicitly recognized the place of inference in science (which presumably includes historical inferences): "Granted that Reason is prior to matter and that the light of primal Reason illuminates finite minds, I can understand how men should come, by observation and inference, to know a lot about the universe they live in."[56]

Further, as we saw above (point 1), Lewis was well aware of the subtle factors that go into the human activity of scientific theorizing. So why did he give such a limited description of "science"? Certainly one factor could be the argumentative context of the essay, for which more subtlety would only be a distraction. But I suspect that the decisive factor is the different kinds of logical reasoning involved. In several places Lewis owned the adage of Aristotle's *Nicomachean Ethics*: "It is the mark of an educated mind to expect that amount of exactness in each kind which the nature of the particular subject admits. It is equally unreasonable to accept merely probable conclusions from a mathematician and to demand strict demonstration from an orator."[57]

Lewis's application was this: "The questions in which mathematicians are interested admit of treatment by a particularly clear and strict technique. Those of the scientist have their own technique, which is not quite the same. Those of the historian and the judge are different again."[58]

Lewis is surely right to see the differences, but it is probably misleading to leave historical inferences outside of "science," especially since the question of what has come to be called "methodological naturalism" in science has become prominent. One author has called in Lewis to support "methodological naturalism" in all science, without recognizing that he has cited one of Lewis's simplified nomothetic descriptions, and that the issue of historical inferences is logically separate.[59] If we must ask Lewis what he would have said about explaining all *events* purely in terms of natural laws, we should attend to his expressed opinion: "In the whole history of the universe the laws of Nature have never produced a single event. They are the pattern to which every event must conform, provided only that it can be induced to happen."[60]

The matter of what might "induce" events is crucial. To those people, theist and non-theist alike, who insist that "scientific" study requires that we always prefer even improbable naturalistic explanations for events rather than say anything beyond the natural has been involved, Lewis replied: "Such a procedure is, from the purely historical point of view, sheer midsummer madness *unless* we start by knowing that any Miracle whatever is more improbable than the most improbable natural event. Do we know this?"[61] Again, any claim to be good science is subject to review for its adherence to sound critical thinking.

It is sometimes asserted that miracles and science cannot dwell together, since appeal to "miracle" is a "science-stopper." In order to discuss this assertion, we need clear definitions of both "science" and "miracle." The discussion above gives some idea of how Lewis might have defined "science"; but what about "miracle"? In choosing a definition of "miracle" for his book of that title, Lewis aimed to be provocative, more than strictly technical: "I use the word Miracle to mean an interference with Nature by supernatural power."[62] He was more technical, and more careful, in his *Studies in Words*, where he explains: "When any agent is empowered by God to do that of which its own *kind* or *nature* would never have made it capable, it is said to act *super-naturally*, above its *nature*."[63] Using this latter definition, we must agree with Lewis, first that studying the regularities ("science") cannot tell us whether an irregularity might occur; second, that the existence of an irregularity, when it is "fed into" the system of nature, in no way undermines the regularities; and third, that if science is to be good it must be subject to the rules of critical thinking, and these rules do not entitle us to rule a miracle out before we consider the evidence for it.

Further, Lewis is clear that good science can help us to discern a miracle, disagreeing with those who "seem to have an idea that belief in miracles arose at a period when men were so ignorant of the cause of nature that they did not perceive a miracle to be contrary to it."[64] On the contrary, he explained, a good judgment of miracle depends on *knowledge* of the things involved—and even partial knowledge is good enough: "No doubt the modern gynaecologist knows several things about birth and begetting which St. Joseph did not know. But those things do not concern the main point that a virgin birth is contrary to the course of nature. And St. Joseph obviously knew *that*." Thus, "Belief in miracles, far from depending on an ignorance of the laws of nature, is only possible in so far as those laws are known."[65]

Lewis and Intelligent Design

LEWIS DIED long before the contemporary movement of "intelligent design" was born. We might suppose that he has no comment on it; at least one author has cited him as a de facto critic of this movement.[66] The raw materials for a "broadly Lewisian" assessment of this movement are nevertheless available, and they show that, so long as the proponents of intelligent design carefully observe some restrictions, intelligent design has promise for giving a credible account of at least some aspects of how we came to be.

On the one hand, though Lewis never discusses William Paley's *Natural Theology* by name,[67] he does mention Paley a few times, and never with approval. He took Paley to be one of those "terrible theologians" who held that "God did not command certain things because they are right, but certain things are right because God commanded them."[68]

Further, Lewis rejects what we may call the Paleyesque plan of inferring a wide range of God's attributes from our observation of nature:

> There is, to be sure, one glaringly obvious ground for denying that any moral purpose at all is operative in the universe: namely, the actual course of events in all its wasteful cruelty and apparent indifference, or hostility, to life.
> At all times, then, an inference from the course of events in this world to the goodness and wisdom of the Creator would have been equally preposterous; and it was never made.[69]

In general, Lewis did not find the large-scale arguments, such as the "cosmological argument," personally helpful, and they play no role in his overall apologetic or piety.[70]

On the other hand, although some critics of "intelligent design" consider it to be a renewed version of Paley's natural theology, this criticism fails to account for the ways in which members of the intelligent design movement have distanced themselves from these aspects of Paley's views: Michael Behe has done so, to name no others.[71]

To say that Paley made mistakes does not establish that therefore *all* kinds of design inference from nature are wrong-headed, and Lewis does *not* fall into that trap. As we will see later in this chapter, Lewis clearly (and persuasively) rejected a naturalistic story for how life originated, and especially for how human reason came about. In other words, his acceptance of "evolution" does not exclude that some events within this evolution include a causal component that goes beyond the natural properties of the things involved. As he wrote: "On any view, the first beginning [of life and of civilization] must have been outside the ordinary processes of nature... Is it not equally reasonable to look outside nature for the real Originator of the natural order?"[72]

3. Lexical Precision

C. S. Lewis was aware that all words have more than one meaning, and that, if we want to read an author sympathetically, we must discern which of these meanings the author intended. This was one of the central points of Lewis's book *Studies in Words*, which articulates several principles of practical lexicography and then illustrates these principles with ten examples, with this end of sympathetic reading in view.

It is not only sympathetic reading that requires lexical precision. So does all clear thinking—that is, we must beware of what logicians call "the fallacy of equivocation." Thus, when Lewis discusses the sin of "pride" in *Mere Christianity* he distinguishes between the innocent things that English speakers might call "pride" (such as pleasure in being praised; warm-hearted admiration for a son, or father, or school, or regiment) and the actual sinful attitude that moralists mean when they speak of pride.[73] Similar terminological distinctions run throughout his writings:

- "In actual modern English usage the verb 'believe,' except for two special usages, generally expresses a very weak degree of opinion."[74]
- "In ordinary usage the word *impossible* generally implies a suppressed clause beginning with the word *unless*."[75]
- "Strictly speaking, there is, I confess, no such thing as 'modern science'. There are only particular sciences, all in a stage of rapid change, and sometimes inconsistent with one another."[76]
- "We must remind ourselves that the word *History* has several senses."[77]

A showcase example of Lewis's carefulness would be in how he treats the word "evolution." Observe how Lewis makes a crucial distinction in lexical usage: "Even, however, if Evolution in the strict biological sense has some better grounds than Professor Watson suggests—and I can't help thinking it must—we should distinguish Evolution in this strict sense from what may be called the universal evolutionism of modern thought."[78]

Lewis commonly distinguished between what he called "evolution as a theorem in biology" and "evolution in the popular imagination (as fueled by the science popularizers)."[79] The two are not the same.

Consider how helpful this kind of precision would be in assessing a fairly recent document, in which the (American) National Science Teachers Association (NSTA) makes a number of assertions about "evolution":

> Evolution in the broadest sense can be defined as the idea that the universe has a history: that change through time has taken place. If we look today at the galaxies, stars, the planet Earth, and the life on planet Earth, we see that things today are different from what they were in the past: galaxies, stars, planets, and life forms have evolved. Biological evolution refers to the scientific theory that living things share ancestors from which they have diverged; it is called "descent with modification." There is abundant and consistent evidence from astronomy, physics, biochemistry, geochronology, geology, biology, anthropology, and other sciences that evolution has taken place.
>
> As such, evolution is a unifying concept for science. The *National Science Education Standards* recognizes that conceptual schemes such as evolution "unify science disciplines and provide students with powerful ideas to help them understand the natural world" (p. 104) and recommends evolution as one such scheme.[80]

To use the same word, "evolution," for these diverse processes is fine, if all we intend is to stress that change has taken place over time. But this masks the (potential) differences between the kinds of processes that may be involved—unless the intention is actually to imply that the processes are quite similar, which should be controversial.

Another teachers' advocacy group, the National Association of Biology Teachers (NABT), does go further in what they insist on in biological evolution: "Evolutionary biology rests on the same scientific methodologies the rest of science uses, appealing only to natural events and processes to describe and explain phenomena in the natural world."[81] But, as we will see below, Lewis was willing to accept evolution in all these realms, including biology, so long as the scientific theory does not require that people limit themselves "only to natural events and processes" for explanation; indeed, for the origin of humankind he would call that "absurd."[82] The difficulty is in the variety of meanings we may attach to "evolution."[83]

One of Lewis's most helpful lexical contributions is his category of "tactical definitions," which are "attempts to appropriate for one side, and deny to the other, a potent word."[84] This is not so much a matter of lexicography as it is of rhetoric. In this process, "[t]he pretty word has to be narrowed down *ad hoc* so as to exclude something he dislikes. The ugly word has to be extended *ad hoc*, or more probably *ad hunc*, so as to bespatter some enemy." An obvious power word in Western cultures is "science"; the term carries with it the prestige of the manifest advances in health care, sanitation, and communication technology that we enjoy, so much so that to call something "unscientific" is to relegate it to the category of "merely of private interest."

4. Ability to Make Fine Distinctions

Lewis's philosophical and literary training enabled him to make fine distinctions, say between "science" and "scientism," or between "evolution" and "evolutionism." He even recognized that the umbrella word "science" can often be misleading, as if all science speaks with one voice. As he observed: "Strictly speaking, there is, I confess, no such thing as 'modern science'. There are only particular sciences, all in a stage of rapid change, and sometimes inconsistent with one another."[85]

This is extremely helpful when we encounter a modern science popularizer, such as John Gribbin, saying the following: "Both evolution and the Big Bang (and all the rest) are based on the same principles, and you can't pick and choose which bits of the scientific story you are going to accept."[86] Now, this may or may not be correct; but simply insisting that both are "science" does nothing to prove it. Gribbin has given us a confused assertion.

We may draw this together, in a broadly Lewisian fashion, to the following principle: When we are faced with statements that begin with "Science says," we should immediately ask, "*Which* science?" And then we can move on to see that "a science" doesn't say anything; scien*tists* do.

Then we can ask, "Which scientists? And have they reasoned so well that I should believe them?" This is especially helpful when someone makes a statement on behalf of *all* science; or when an expert in one science (say, physics) tries to speak authoritatively about some other field (say, linguistics, or psychology): just because he is a "scientist" does not mean I am obligated to believe him. As we have already seen, Lewis would have us consider whether he has reasoned well. (Of course, if someone speaks as an expert in his own field, then we ought to pay closer attention!)

5. Sophisticated Textual Exegesis

CERTAINLY IF WE WANT TO address science and religion questions, we have to grapple with the nature of the relevant religious texts: What sorts of things do they talk about, and what may their believing readers reasonably turn to them for? Lewis was both a Christian and a literary scholar, and his approach to the Bible displays both respect for its historical element (including the miraculous) and regard to the different literary types contained in it.[87]

In his *Reflections on the Psalms*, Lewis made it clear that he was not what he called a "Fundamentalist." He did not define the term, but it would appear from the context that he meant someone with "a prior belief that every sentence of the Old Testament has historical or scientific truth."[88] Lewis added: "But this [prior belief] I do not hold, any more than St. Jerome did when he said that Moses described Creation 'after the manner of a popular poet' (as we should say, mythically) or than Calvin did when he doubted whether the story of Job were history or fiction."[89]

Lewis here refers to Jerome, but no one has located the exact spot in which Jerome makes this claim. The closest source is a passage in the English proto-Reformer John Colet (1467–1519), who wrote in a letter to one Radulphus: "Thus Moses arranges his details in such a way as to give the people a clearer notion, and he does this *after the manner of a popular poet*, in order that he may the more adapt himself to the spirit of simple rusticity, picturing a succession of things, works, and times, of such a kind as there certainly could not be in the work of *so great a Workman*."[90]

With this principle in mind, Lewis addresses the possibility that the creation story in Genesis is in some way "derived from earlier Semitic stories which were Pagan and mythical"—a view that had become widely spread by his time, and which was held to discredit Genesis. But Lewis shows his good literary and philosophical sense by first insisting, "We must of course be quite clear what 'derived from' means. Stories do not reproduce themselves like mice."[91] He observed that it is *persons* who do the retelling and revising of stories for various ends:

> Thus at every step in what is called—a little misleadingly—the "evolution" of a story, a man, all he is and all his attitudes, are involved. And no good work is done anywhere without aid from the Father of Lights. When a series of such re-tellings turns a creation story which at first had almost no religious or metaphysical significance into a story which achieves the idea of true Creation and of a transcendent Creator (as *Genesis* does), then nothing will make me believe that some of the re-tellers, or some one of them, has not been guided by God.[92]

Hence, although Lewis found much that he deemed poetical or even mythical in the Genesis creation story, he was nevertheless willing to attach to it some kind of referent. For example:

> We read in *Genesis* (2, 7) that God formed man of the dust and breathed life into him. For all the first writer knew of it, this passage might merely illustrate the survival, even in a truly creational story, of the Pagan inability to conceive true Creation, the savage, pictorial tendency to imagine God making things "out of" something as the potter or the carpenter does. Nevertheless, whether by lucky accident or (as I think) by God's guidance, it embodies a profound principle. For in any view man is in one sense made "out of" something else. He is an animal; but an animal called to be, or raised to be, or (if you like) doomed to be, something more than an animal. On the ordinary biological view (what difficulties I have about evolution are not religious) one of the primates is changed so that he becomes man; but he remains still a primate and an animal.[93]

Now, Lewis elsewhere makes it clear that he thought this "changing" of one of the primates to become man was both historical and supernatural.[94] Thus the poetical or pictorial style of the Genesis story does not prevent it from referring to a real event in the history of the world.

Much has happened in the study of the ancient Near East since Lewis's time. While it was once common to suppose that the Babylonian poem *Enuma Elish* was the relevant "ancestor" of the Genesis creation story,[95] and though some Biblical scholars still think this way, Assyriologists now find in Genesis 1–11 a set of "parallels" to some much older Mesopotamian sources.[96] To call them "parallels" highlights the problems connected with derivation; it allows that Genesis is a response to, a comment upon, or even a refutation of, the Mesopotamian stories, without saying what the exact literary relationship is. Indeed, these Mesopotamian stories served to instill a worldview in their society, and Genesis instills an alternative worldview in ancient Israel.

Some may prefer a higher level of literalism in their reading of Genesis than Lewis did, but the broadly Lewisian points still stand: First, the possible effect of "pagan" stories on the origin of Genesis need not detract from its inspiration; and second, the possibility (in my mind, the near assurance) of shared motifs and literary conventions with other ancient near Eastern stories need not detract from the referentiality, nor even from the historicity, of the Genesis material—so long as we do not identify historicity with literalism in interpretation.[97]

6. A Sensible "Freedoms and Limitations" Approach

Francis A. Schaeffer (1912–1984) was an American Presbyterian pastor who founded and operated L'Abri, an apologetics mission in the Swiss Alps oriented to young people. Of interest here is Schaeffer's confidence that ultimately there is no conflict between the testimony of the Bible and the testimony of creation (as studied in the sciences). Of course, there are challenges and disagreements among the *interpreters* of these testimonies, and thus potential conflicts between these interpretations. In his pamphlet *No Final Conflict,* Schaeffer articulated an approach that he called "freedoms and limitations": There is a range of reasonable scenarios by which we may address the apparent conflicts between the Bible and the sciences, and yet there are limits to this range, limits set both by basic Biblical concepts and by good human judgment.[98]

In view of the possible presence of pictorial material and literary conventions in Genesis 1–11, this approach of Schaeffer's is surely the best way for believers to commend their faith in the larger world.

C. S. Lewis had his own version of this approach (developed, of course, entirely independently of Schaeffer). For example, he was happy to entertain a variety of scientific-historical scenarios for the origin of humankind, but none of them would be valid if they were to deny the mystery of reason, or any implications that follow from that mystery. Likewise, these scenarios must not deny the objectivity of ethical judgments.

Consider, for example, his treatment of human origins in the context of his chapter on "The Fall of Man" in *The Problem of Pain*.[99] In discussing the Biblical story of Adam and his sin, and the degree to which that event may have affected the rest of us (traditionally the descendants of Adam and Eve), he acknowledges this about "the Fathers" (the leading theologians in the first few Christian centuries): "Wisely, or foolishly, they believed that we were *really*—and not simply by legal fiction—involved in Adam's action."[100]

Now, Lewis wanted to do justice to this belief, at the same time as he recognized that twentieth-century scientific beliefs made literalism difficult. So he set about addressing, with his characteristic clarity, how these two sets of beliefs might relate to one another. "Many people think that this proposition"—that God created Man good, but then Man fell by disobedience—"has been proved false by modern science," with its view that "men have arisen from brutality and savagery." Lewis thought that this belief showed "complete confusion":

> *Brute* and *savage* both belong to that unfortunate class of words which are sometimes used rhetorically, as terms of reproach, and sometimes scientifically, as terms of description; and the pseudo-scientific argu-

ment against the Fall depends on a confusion between the usages. If by saying that man rose from brutality you mean simply that man is physically descended from animals, I have no objection. But it does not follow that the further back you go the more *brutal*—in the sense of wicked or wretched—you will find man to be.

Lewis concluded that "Science, then, has nothing to say either for or against the doctrine of the Fall," and he proceeded to offer a scenario harmonizing human evolution and the Fall that he thought might be "a not unlikely tale."[101] Lewis argued that over many centuries "God perfected the animal form which was to become the vehicle of humanity and the image of Himself." God then "caused to descend upon this organism, both on its psychology and physiology, a new kind of consciousness… which knew God, which could make judgments of truth, beauty and goodness, and which was so far above time that it could perceive time flowing past." Lewis does not speculate about "how many of these creatures God made, nor how long they continued in the Paradisal state." But he stressed that "sooner or later they fell. Someone or something whispered that they could become as gods… We have no idea in what particular act, or series of acts, the self-contradictory, impossible wish found expression. For all I can see, it might have concerned the literal eating of a fruit, but the question is of no consequence."

It is clear from the context, and from Lewis's other writing, that by "God caused to descend upon this organism, both on its psychology and physiology, a new kind of consciousness" he meant something that was *super*natural; as he said in *Miracles*: "To believe that Nature produced God, or even the human mind, is, as we have seen, absurd."[102] And notice that this "new kind of consciousness" enables the creature to reason about transcendent moral realities, which again cannot be a simple outgrowth of its material capacities.

We might want to revise this scenario in some of its particulars. I have offered my own revision for it elsewhere, and my result is still "broadly Lewisian."[103] The point here is that, according to Lewis, whatever the details of the story we tell about human origins, there are limits to what we may suggest and still be within the bounds of good critical thinking—which, as we have seen, is necessary if we want to be practicing good science.

What applies to the origin of humankind applies to evolutionary theory in general. Lewis mentioned evolution frequently in his writings, generally to distinguish the scientific theory from the ideological extrapolations some made of it (as above). But he said more about the theory itself. A fair sample would be this:

> Again, for the scientist Evolution is a purely biological theorem. It takes over organic life on this planet as a going concern and tries to explain certain changes within that field. It makes no cosmic statements, no metaphysical statements, no eschatological statements... It does not in itself explain the origin of organic life, nor of the variations, nor does it discuss the origin and validity of reason. It may well tell you how the brain, through which reason now operates, arose, but that is a different matter. Still less does it even attempt to tell you how the universe as a whole arose, or what it is, or whither it is tending. But the Myth knows none of these reticences... 'Evolution' (as the Myth understands it) is the formula of *all* existence.[104]

The kind of evolutionary theory that did not bother Lewis theologically "does not in itself explain the origin of organic life, nor of the variations, nor... of reason"; that is, it does not insist beforehand that we may only allow a purely naturalistic scenario for the whole development of life.[105]

Nominally, at least, some leading evolutionary biologists support Lewis on this. For example, D. M. S. Watson (1886–1973), Professor of Zoology and Comparative Anatomy at University College, London (1921–1951), acknowledged: "But whilst the fact of evolution is accepted by every biologist the mode in which it has occurred and the mechanism by which it has been brought about are still disputable."[106] The National Science Teachers Association says something very similar: "There is no longer a debate among scientists about whether evolution has taken place. There is considerable debate about how evolution has taken place:

What are the processes and mechanisms producing change, and what has happened specifically during the history of the universe?"[107]

Even though these statements sensibly refuse to decide ahead of time what kinds of factors can be involved, other statements, as we have seen, are emphatic in ruling some things out. The National Association of Biology Teachers insists: "Evolutionary biology rests on the same scientific methodologies the rest of science uses, appealing only to natural events and processes to describe and explain phenomena in the natural world. Science teachers must reject calls to account for the diversity of life or describe the mechanisms of evolution by invoking non-naturalistic or supernatural notions."[108]

I do not suggest that Lewis thought that "non-naturalistic" notions *must* be involved in any extensive fashion in evolution, nor whether any of these would be readily perceptible to human students, nor even whether it was appropriate for him to have much of an opinion. The clear exception is the origin and development of humankind.

Consider some more of the features that Lewis touches on, which distinguish humans from the rest of the animals. The motivation for science itself is distinctively human: "One of the things that distinguishes man from the other animals is that he wants to know things, wants to find out what reality is like, simply for the sake of knowing."[109] Human friendship is another feature that resists explanation purely in terms of natural development of animal capacities: "Friendship is—in a sense not at all derogatory to it—the least *natural* of loves; the least instinctive, organic, biological, gregarious and necessary... We can live and breed without Friendship. The species, biologically considered, has no need of it... [Friendship] has no survival value; rather it is one of those things which give value to survival."[110]

Some researchers have indeed tried to argue that human friendship is continuous with some aspects of animal behavior, but they must rest their argument on defining friendship down. A survey article in *Science News*, entitled "Beast buddies," concludes: "Harder to understand though, according to Silk, are the bonds so close and widespread in *Homo sapiens*. She says, 'None of our models of reciprocity [among nonhuman animals] can accommodate the psychology of human friendship.'"[111]

Lewis would have us believe that there is something fundamentally unreasonable in the insistence (such as the NABT has apparently made) that even humankind arose by a purely natural process; and since it is unreasonable, it is bad scientific history. Lewis had undoubtedly met people who made just such an insistence, even in the name of "science." Of them he said:

> They ask me at the same moment to accept a conclusion and to discredit the only testimony on which the conclusion can be based. The difficulty is to me a fatal one; and the fact that when you put it to many scientists, far from having an answer, they seem not even to understand what the difficulty is, assures me that I have not found a mare's nest but detected a radical disease in their whole mode of thought from the very beginning.[112]

To see how Lewis's approach helps us to face contemporary challenges, consider the following example. Anthony Cashmore is a professor of biology at the University of Pennsylvania, specializing in "the mechanism by which plants respond to light."[113] He was elected to the National Academy of Sciences in 2003, and the *Proceedings of the National Academy of Sciences* published his inaugural article, "The Lucretian swerve: The biological basis of human behavior and the criminal justice system," in 2010.[114] Cashmore aims to show that, since human behavior is the product of genes, environment, and "stochastic" factors (that is, they are probabilistic), therefore there is no such thing as free will. This in turn means that "individuals cannot logically be held responsible for their behavior"; which then leads to Cashmore's purpose for writing, namely a proposal to reform the American criminal justice system.

In making the argument about how the biological factors determine behavior, Cashmore does cite a few studies in cognitive science, regarding the relationship between measurable brain activity and human choices.

He does not claim that anyone actually understands the brain processes, or what consciousness is; he rather expects that at some point we will have a full explanation for how it arises from the chemical properties of the nervous system. He nevertheless insists, "as living systems we are nothing more than a bag of chemicals"; "not only do we have no more free will than a fly or a bacterium, in actuality we have no more free will than a bowl of sugar. The laws of nature are uniform throughout, and these laws do not accommodate the concept of free will."

Because "progress in understanding the chemical basis of behavior will make it increasingly untenable to retain a belief in the concept of free will," therefore "it is time for the legal system to confront this reality." All of the reforms he proposes stem from "the elimination of the illogical concept that individuals are in control of their behavior in a manner that is something other than a reflection of their genetic makeup and their environmental history," and this will "hopefully minimize the retributive aspect of criminal law." Persons convicted of crimes will then be given the appropriate psychiatric help (as specified by a "court-appointed panel of experts").

There is much to say about the overall logic of Cashmore's argument, and many of the details as well. For now I will content myself with observing that Lewis would surely point out that Cashmore, in declaring our thoughts to be merely a biological phenomenon, has undercut anyone's right to believe such a claim. Cashmore seems to have taken it as fundamental to science that we must seek purely material and, apparently, reductionistic, explanations for everything.

Lewis, who wrote, "A man's rational thinking is *just so much* of his share in eternal Reason as the state of his brain allows to become operative,"[115] would not be surprised at the close connection between brain activity and decisions, although he would likely also point out how little these findings actually do explain. He would surely also note what problems we make for ourselves if we suppose that it is even reasonable to posit that chemical events in the brain, strictly speaking, *cause* thoughts or choices.

Further, we recognize that free will and moral responsibility are parts of a larger realm of discourse, in which there is some transcendent norm that we are obligated to comply with. And Cashmore has not evaded such transcendence. Why does he not suggest that we simply eliminate those who commit crimes—whether from the population, or at least from the gene pool (say, by sterilization)? Surely it is because he sees as clearly as anyone else that we *should not* do such horrors to our fellow humans. Why does he think we ought to restructure our criminal justice system? Is it not because he considers it *unjust* to punish people for things for which they are not responsible? And who gets to choose these panels of experts, what kind of decision-making does he expect from them, and to whom will they be accountable? Will they be "wrong" if they accept bribes or other favors? I must believe that Cashmore expects everyone to behave with the utmost honesty and fairness, and that they are blameworthy if they do not. And even to make the case for something, to try to persuade us that it is *true*, is in effect to insist that the readers *ought to* believe it, and are culpable if they do not.[116]

It is to Cashmore's credit that he commits these inconsistencies: They are his humanity breaking through. As Lewis put it: "Holding a philosophy which excludes humanity, they yet remain human... They know far better than they think they know."[117]

We might suspect that Lewis would add, along with his Professor Digory Kirke from his Narnian Chronicles, "Logic! Why don't they teach logic at these schools?"[118]

7. Empathy with Those Criticizing Religion

Lewis tells us that he was once an atheist, or perhaps agnostic. His process of conversion took him from skepticism, to philosophical idealism, to theism, to full-fledged Christian belief. He became a convinced theist in 1929 when he was already teaching at Oxford; and he became a convinced Christian in 1931, after a long evening talk with his fellow scholars J. R. R. Tolkien and Hugo Dyson. Further, he had suffered many things: He lost his mother early in his childhood, he was estranged from his father, he had unhappy experiences at the boarding schools his father chose, he was wounded in World War I, and had a difficult domestic life with Mrs. Moore, the mother of his friend who was killed in the War.[119]

All of this meant that Lewis could sympathize with those for whom he wrote: with their intellectual hurdles to Christian belief, and with their personal experiences of grief. As he poignantly put it, "I was at this time [in boarding school] living, like so many Atheists or Antitheists, in a whirl of contradictions. I maintained that God did not exist. I was also very angry with God for not existing. I was equally angry with Him for creating a world."[120] One gets the impression that Lewis's arguments derive, not only from his earnest conversations with his fellow academics, and with the military personnel whom he spoke with during World War II, but also from the struggles of his own heart.

Lewis knew that what leads most people to believe in, or to reject belief in, God, is only partly intellectual. At the same time, he understood the profound importance of cultivating a climate that can remove barriers to belief, whether those barriers are inspired by a misguided view of science or something else. As Lewis's friend Austin Farrer, a theologian and clergyman who ministered to Lewis in his final illness, put it, "[f]or though argument does not create conviction, the lack of it destroys belief. What seems to be proved may not be embraced; but what no one shows the ability to defend is quickly abandoned. Rational argument does not create belief, but it maintains a climate in which belief may flourish."[121]

Finally, Lewis showed exemplary humility in what he thought his apologetic could achieve: "I have found that nothing is more dangerous to one's own faith than the work of an apologist. No doctrine of that Faith seems to me so spectral, so unreal as one that I have just successfully defended in a public debate."[122] At the same time, he has modeled a mature faith that robustly engages the challenges we face, and offered invaluable tools for commending that faith to the world at large.

Endnotes

1. Cited from Walter Hooper, "Preface," in Lewis, *Selected Literary Essays*, edited by Walter Hooper (Cambridge: Cambridge University Press, 1969), vii–xx, at xiii. George Sayer gives some more detail in his biography, *Jack: A Life of C. S. Lewis* (Wheaton, IL: Crossway, 1994), 176–83. Lewis had earned three first-class undergraduate degrees at University College: "Honour Mods (Greek and Latin texts), Greats (classical philosophy), and English Language and Literature." Sayer, *Jack*, 175. Austin Farrer points out that this meant that Lewis was familiar with professional philosophical trends in the earlier part of the twentieth century, and less so with those from later: "The Christian apologist," in Joceyln Gibb, editor, *Light on C. S. Lewis* (London: Geoffrey Bles, 1965), 23–43, at 30–31.

2. A very small sampling includes Steven Jon James Lovell, *Philosophical Themes from C. S. Lewis* (University of Sheffield PhD, 2003); Richard Purtill, *C. S. Lewis' Case for the Christian Faith* (San Francisco: Ignatius, 2004 [1985]); J. R. Lucas, "Restoration of Man: A Lecture given in Durham on Thursday October 22, 1992," accessed May 3, 2012, http://users.ox.ac.uk/~jrlucas/lewis.html; Houston A. Craighead, "C. S. Lewis' teleological argument," *Encounter* 57, no. 2 (1996): 171–85; Victor Reppert, *C. S. Lewis's Dangerous Idea: A philosophical defense of Lewis's Argument from Reason* (Downers Grove, IL: InterVarsity Press, 2003).

3. For one example among many: N. T. Wright, "Simply Lewis: Reflections on a master apologist after 60 years," *Touchstone* 20, no. 2 (2007). Further, the English-Canadian Anglican theologian J. I. Packer, with whom I worked on the *English Standard Version*, has recounted to me some of his memories of Lewis at Oxford in the 1940s, always speaking with deep appreciation.

4. See Lewis's recognition of this situation in "Dogma and the Universe" (1943) in C. S. Lewis, *God in the Dock: Essays on Theology and Ethics*, edited by Walter Hooper (Grand Rapids: Eerdmans, 1970), 38. Since many of Lewis's works appear in several editions, with different page numbers (and some of these are online), and since the collections of essays gather things written at different times, my convention for citing Lewis is the following: the books I will cite from the edition I own, but also with the chapter or section number; the essays I will cite from the collection, but also with the title and (probable) date.

5. See "Myth Became Fact" (1944) in Lewis, *God in the Dock*, 63.

6. Roger Lancelyn Green and Walter Hooper, *C. S. Lewis: A Biography* (Glasgow: William Collins, 1974), 229.

7. Examples include the correspondence of the English philosopher Mary Midgley with Arend Smilde of the Netherlands, entitled "I still find his [Lewis's] thought terribly helpful," http://www.lewisiana.nl/marymidgley/; see also Antony Flew's comments in Gary R. Habermas and Antony Flew, "From Atheism to Deism: A Conversation between Antony Flew and Gary R. Habermas," in David Baggett, Gary R. Habermas and Jerry L. Walls, editors, *C. S. Lewis as Philosopher: Truth, Goodness, and Beauty* (Downers Grove, IL: InterVarsity, 2008), 37–52.

8. Francis Collins, *The Language of God: A Scientist Presents Evidence for Belief* (New York: Free Press, 2006).

9. For example, a serialized article by the philosopher Michael Peterson argues that Lewis offers a trenchant critique of the ideas behind what is now known as "intelligent design";

Peterson's published version is "C. S. Lewis on Evolution and Intelligent Design," *Perspectives on Science and Christian Faith* 62, no. 4 (2010): 253–266. I judge Peterson to have presented an inadequate appraisal of both Lewis and intelligent design.

10. See, for example, C. John Collins, *The God of Miracles: An Exegetical Examination of God's Action in the World* (Wheaton, IL: Crossway, 2000); *Science and Faith: Friends or Foes?* (Wheaton, IL: Crossway, 2003).

11. My dissertation is called *Homonymous Verbs in Biblical Hebrew: An Investigation of the Role of Comparative Philology* (University of Liverpool PhD, 1988).

12. In the same way, the linguist-logician Arthur Gibson, in *Biblical Semantic Logic* (New York: St. Martin's Press, 1981), commends a study by A. R. Millard: "although it is short and does not employ formal analysis, it does not commit the aforementioned errors. Hence a formal analysis is not necessary to avoid Barr's criticisms." Gibson, *Biblical Semantic Logic*, 5, n. 1.

13. Although I have long thought along these lines, I attribute the delightful term "broadly Lewisian" to Steven Lovell's *Philosophical Themes from C. S. Lewis*, e.g., 10.

14. Richard Purtill, *C. S. Lewis' Case for the Christian Faith*, gives an excellent list of factors that contributed to Lewis's success, in the intellectual, imaginative, and moral realms. My list complements his.

15. J. B. S. Haldane, *Possible Worlds and Other Essays* (New Brunswick, NJ: Transaction Publishers, 2002 [1927]), 7.

16. C. S. Lewis, *The Discarded Image: An Introduction to Medieval and Renaissance Literature* (Cambridge: Cambridge University Press, 1964), 22.

17. Ibid., 97–99.

18. For example, a quick sampling from Lewis's *God in the Dock*: "Dogma and the Universe" (1943), 39; "Religion and Science" (1945), 74; "Christian Apologetics" (1945), 99; also Lewis's *Miracles: A Preliminary Study* (New York: Macmillan, 1960), ch. 7, 49. See further Lewis's previously unpublished "Reply to Professor Haldane" in *Of Other Worlds: Essays and Stories*, edited by Walter Hooper (New York: Harcourt Brace Jovanovich, 1966), 75. My hunch is that Haldane's work lies behind a number of Lewis's writings, but I will not pursue that here. Walter Hooper indicates that Haldane "hated Christianity intensely"; and once when Haldane addressed the Oxford Socratic club, he avoided Lewis's response to his paper: "Professor Haldane clearly had no desire to confront Lewis and the other Socratics, and he ended his short reply to Mr. Crombie [his respondent] 'with an impressive running panegyric of atheism of which the last word was perfectly timed to coincide with his exit'." See Hooper, "Oxford's bonny fighter," in James T. Como, editor, *Remembering C. S. Lewis: Reflections of Those Who Knew Him* (San Francisco: Ignatius, 2005), 241–308, at 280–81. Another popular writer whom Lewis opposed, perhaps more influential but less intellectually weighty, was H. G. Wells. For discussion of Lewis's engagement with Wells, see Thomas C. Peters, "The War of the Worldviews: H. G. Wells and Scientism versus C. S. Lewis and Christianity," in David Mills, editor, *The Pilgrim's Guide: C. S. Lewis and the Art of Witness* (Grand Rapids: Eerdmans, 1998), 203–20.

19. Eoin O'Carroll, "Nicholas Steno: The Saint who Undermined Creationism," *Christian Science Monitor*, January 11, 2012 (online edition).

20. Jay Richards, "No, Copernicus did not remove us from the hub of the universe," *Evolution News and Views*, January 13, 2012, accessed May 3, 2012, http://www.evolution-news.org/2012/01/copernicus_did_055071.html.
21. As Lovell put it, "Kuhn and other philosophers of science may express their positions with greater sophistication, but once again, the strength of Lewis' presentation is his clarity." Lovell, *Philosophical Themes from C. S. Lewis*, 8.
22. Lewis, *Discarded Image*, 219–220. For a very similar take on the history, see Michael Polanyi, *Personal Knowledge: Towards a Post-Critical Philosophy* (Chicago: University of Chicago Press, 1958), 3–4.
23. Lewis, *Discarded Image*, 222; for a similar point, see Haldane, *Possible Worlds*, 225–29 (e.g., "the experience of the past makes it clear that many of our most cherished scientific theories contain so much falsehood as to deserve the title of myths," 228).
24. C. S. Lewis, *Surprised by Joy: The Shape of My Early Life* (London: Collins, 1959), ch. 13, 167.
25. Note how David Lindberg, *The Beginnings of Western Science* (Chicago: University of Chicago Press, 1992), assessed Aristotle: "The proper measure of a philosophical system is not the degree to which it anticipated modern thought, but its degree of success in treating the philosophical problems of its own day. If a comparison is to be made, it must be between Aristotle and his predecessors, not Aristotle and the present. Judged by such criteria, Aristotle's philosophy is an astonishing achievement." Lindberg, *Beginnings*, 67.
26. Lewis, *Discarded Image*, 92.
27. Ibid., 92–93.
28. Lewis, "*De descriptione temporum*" (1955) in Lewis, *Selected Literary Essays*, 7.
29. Lewis, *The Abolition of Man* (New York: Touchstone, 1996). This book is online at http://www.columbia.edu/cu/augustine/arch/lewis/abolition1.htm.
30. Lewis, *Abolition*, ch. 1, 31.
31. Ibid., ch. 3, 65.
32. Ibid., ch. 3, 82.
33. Ibid., ch. 3, 83.
34. As M. D. Aeschliman observed in "C. S. Lewis on Mere Science," *First Things* (October 1998): 16–18, "Lewis knew that science was one of the great products of the human mind, but insisted that it was a subset of reason and not simply equivalent to it."
35. Lewis, "Is Progress Possible? Willing Slaves of the Welfare State" (1958) in Lewis, *God in the Dock*, 315. Lewis was a long-time opponent of vivisection in scientific research: "The victory of vivisection marks a great advance in the triumph of ruthless, non-moral utilitarianism over the old world of ethical law." Lewis, "Vivisection" (1948), *God in the Dock*, 228.
36. For a definition of these positions, see Lewis, *Miracles*, ch. 2, 6: "What the Naturalist believes is that the ultimate Fact, the thing you can't go behind, is a vast process in space and time which is *going on of its own accord*." The materialist believes this, and insists that everything is governed by purely material causes. Lewis seems to see these as alternative names for the position he opposes in *Miracles*. See also Aeschliman, "C. S. Lewis on Mere Science," 16.

37. Lewis, *Miracles*, ch. 5, 25.
38. Quoted by Lewis in *Miracles*, ch. 3, 15, citing Haldane's *Possible Worlds*, 209. To be sure, Haldane himself did not draw precisely the conclusions that Lewis did regarding the special origin of human consciousness in biological history: "But consciousness arises anew in every human being. Its first origin on the earth presents no more and no less mystery than its last." Haldane, *Possible Worlds*, 44. Surely this underestimates the problem! The second edition of Lewis's *Miracles* included a revised version of this argument, to account of the critique offered by the Roman Catholic philosopher G. E. M. Anscombe in her paper for the Oxford Socratic Club, February 2, 1948; Miss Anscombe is reported to have found the revision an improvement. See, for example, Reppert, *Lewis's Dangerous Idea*; Reppert, "Defending the Dangerous Idea," in *C. S. Lewis as Philosopher*, 53–67; John Dolan, "G. E. M. Anscombe: Living the Truth," *First Things* (May 2001): 11–13. See also Angus Menuge, *Agents under Fire: Materialism and the Rationality of Science* (Lanham, MD: Rowman & Littlefield, 2004). Those who think that Miss Anscombe's critique actually made Lewis abandon this line of argument are hard-put to account for exactly the revised version, in short form, in Lewis's lectures, *Discarded Image*, 166 (published in 1964).
39. For the sake of space I have not included Lewis's own insistence that reasoning about moral questions, as well as about science, requires transcendence; see *Miracles*, ch. 5, 35; "On Ethics," "*De Futilitate*," "The Poison of Subjectivism," *Christian Reflections*, 44–81.
40. Lewis, *Miracles*, ch. 13, 105–106.
41. Lewis, "The Funeral of a Great Myth" (ca. 1945), *Christian Reflections*, 83; similarly, Lewis, "The World's Last Night" (1952), in Lewis, *The World's Last Night and Other Essays* (New York: Harcourt Brace Jovanovich, 1960), 101. For a parallel, though brief, critique of how Freudians were extrapolating from the "actual medical theories and techniques of the psychoanalysts" to the "general philosophical view of the world which Freud and some others have gone on to add to this," see Lewis, *Mere Christianity* (New York: Macmillan, 1952), Bk. III, ch. 4, 83–84.
42. Lewis, "Religion and Rocketry" (1958), *The World's Last Night*, 84.
43. Lewis, "Christian Apologetics" (1945), *God in the Dock*, 92–93.
44. Ibid.
45. Lewis, "Dogma and the Universe" (1943), *God in the Dock*, 39, On this very page, and in many other places, Lewis cites Eddington's popularizing treatments of science; and for Eddington's development of the model of George Lemaître, which became the Big Bang idea, see John North, *The Norton History of Astronomy and Cosmology* (New York: Norton, 1995), 524–530.
46. Lewis, "Reply to Professor Haldane" (ca. 1946), *Of Other Worlds*, 76–77.
47. Lewis, "Is Theology Poetry?" (1944), in *The Weight of Glory and Other Addresses*, edited by Walter Hooper (New York: Simon & Schuster, 1996 [1980]), 96.
48. Ibid., 94; see also p. 105 for Lewis's positive view of honest science (quoted below).
49. J. B. S. Haldane, "Auld Hornie, F. R. S.," *Modern Quarterly* (Autumn 1946); available online at http://marxists.org/archive/haldane/works/1940s/oncslewis.htm (accessed May 3, 2012).
50. Lewis, "Reply to Professor Haldane" (ca. 1946), *Of Other Worlds*, 77.
51. Ibid., 78.

52. Ibid., 80.
53. For a horrifying account (just one of many), see "Thousands Sterilized, a State Weighs Restitution," *New York Times*, December 9, 2011, accessed May 3, 2012, http://www.nytimes.com/2011/12/10/us/redress-weighed-for-forced-sterilizations-in-north-carolina.html.
54. Lewis, "Historicism" (1950), *Christian Reflections*, 112.
55. Ian Barbour, *Religion in an Age of Science* (New York: HarperCollins, 1990), 66–71.
56. Lewis, "Is Theology Poetry?" (1944), *Weight of Glory*, 105.
57. Aristotle, *Nicomachean Ethics*, 1094b, ll. 23–25 (I.iii.4), from the Loeb edition (1934; trans. H. Rackham). Lewis acknowledges this principle (not always by name) in, e.g., "On Obstinacy in Belief" (1955), *World's Last Night*, 20; "Historicism" (1950), *Christian Reflections*, 101; *Discarded Image*, 78. Lewis certainly valued this Aristotelian work highly, as shown in his chapter in *The Four Loves* (London: Geoffrey Bles, 1960) on "Friendship"; and in his appropriation of Aristotle's description of character formation (on which see discussion in my *God of Miracles*, 104–106).
58. "On Obstinacy in Belief" (1955), *World's Last Night*, 20.
59. Peterson, "C. S. Lewis on Evolution and Intelligent Design," 257. He cites a passage from *Mere Christianity*, Book I, ch. 4: this occurs in a radio broadcast talk, and abbreviation is a strategic necessity, which means that the cited passage is poorly chosen.
60. Lewis, "The Laws of Nature" (1945), *God in the Dock*, 77.
61. Lewis, *Miracles*, ch. 13, 101.
62. Ibid., ch. 2, 5.
63. Lewis, *Studies in Words* (Cambridge: Cambridge University Press, 1967), 64, which gives the definition that is in keeping with the conclusions in my *God of Miracles*. For more discussion see Alvin Plantinga, "Methodological naturalism? Part 2," *Origins and Design* 18, no. 2 (1997), accessed May 3, 2012, http://www.arn.org/docs/odesign/od182/methnat182.htm.
64. Lewis, Miracles, ch. 7, 46.
65. Ibid., 46–47; see also my "Miracles, Intelligent Design, and God-of-the-Gaps," *Perspectives on Science and Christian Faith*, 55, vol. 1 (2003): 22–29.
66. Peterson, "C. S. Lewis on Evolution and Intelligent Design."
67. William Paley, *Natural Theology: Or, evidence of the existence and attributes of the Deity, collected from the appearances of nature* (Oxford: Oxford University Press, 2006 [1802]).
68. Lewis, *Reflections on the Psalms*, ch. 6, 61. Although he does not cite Paley here, he does in *The Problem of Pain* (New York: Macmillan, 1962), ch. 6, 100.
69. The first quotation comes from Lewis, "De Futilitate" (ca. 1943), *Christian Reflections*, 69; the second comes from Lewis, *Problem of Pain*, ch. 1, 16.
70. In a letter to Dom Bede Griffiths (a former student who became a Benedictine monk in India), Lewis wrote, "The Cosmological argument is, for some people at some times, ineffective. It always has been for me." Letter of May 28, 1952, cited in Richard Purtill, *C. S. Lewis' Case for the Christian Faith*, 33.

71. Michael Behe, *Darwin's Black Box: The Biochemical Challenge to Evolution* (New York: Free Press, 1996), 211–16. I have given my own comments in this regard in my "Miracles, Intelligent Design, and God-of-the-Gaps," 24–25.
72. Lewis, "Two Lectures" (1945), *God in the Dock*, 211. Purtill, *C. S. Lewis' Case for the Christian Faith*, 39–45, considers Lewis's treatment of the origin of human reason to be a unique and creative version of the argument from design.
73. Lewis, *Mere Christianity*, Book III, ch. 8, 108–114.
74. Lewis, "On Obstinacy in Belief" (1955), *The World's Last Night*, 14.
75. Lewis, *The Problem of Pain*, ch. 2, 27.
76. Lewis, "The Funeral of a Great Myth" (ca. 1945), *Christian Reflections*, 82.
77. Lewis, "Historicism" (1950), *Christian Reflections*, 105.
78. Lewis, "Is Theology Poetry?" (1944), *Weight of Glory*, 104. The Professor Watson here is D. M. S. Watson (1886–1973), of London; although he was cited in an essay "Science and the BBC," in *Nineteenth Century* (April 1943), the actual quotation is a pastiche from his article "Adaptation," in the *British Association for the Advancement of Science, Report of the Ninety-Seventh Meeting* (London: Office of the British Association, 1929), 88–99: "Evolution itself is accepted by zoologists not because it has been observed to occur or is supported by logically coherent arguments, but because it does fit all the facts of Taxonomy, of Palaeontology, and of Geographical Distribution, and because no alternative explanation is credible" (88); and "the Theory of Evolution itself, a theory universally accepted, not because it can be proved by logically coherent evidence to be true, but because the only alternative, special creation, is clearly incredible" (95).
79. Lewis frequently used the word "theorem" for the biological view of evolution; I follow his example (and his veneration for the *Oxford English Dictionary*, finished in 1933) in noting that he must mean "a universal or general proposition or statement, not self-evident (thus distinguished from an axiom) but demonstrable by argument," but under sense b (the general sense, rather than the specific sense of mathematical demonstration).
80. "NSTA Position Statement: The Teaching of Evolution," adopted July 2003, accessed December 2011, http://www.nsta.org/about/positions/evolution.aspx.
81. National Association of Biology Teachers, "NABT Position Statement on Teaching Evolution" (2011), accessed December 2011, http://www.nabt.org/websites/institution/index.php?p=92.
82. Lewis, *Miracles*, ch. 4, 32.
83. A valuable survey, in keeping with the Lewisian spirit, of how the different meanings of the word "evolution" can lead to confusion is Stephen C. Meyer and Michael Newton Keas, "The Meanings of Evolution," in John Angus Campbell and Stephen C. Meyer, editors, *Darwinism, Design, and Public Education* (East Lansing, MI: Michigan State University Press, 2003), 135–156. In my *Science and Faith*, 335–337, I observe how a leading Darwinism advocate, Eugenie Scott, falls foul of the fallacy of equivocation.
84. Lewis, *Studies in Words*, 17–19.
85. Lewis, "The Funeral of a Great Myth" (ca. 1945), *Christian Reflections*, 82.
86. John Gribbin, *Almost Everyone's Guide to Science* (London: Phoenix, 1999), 2.

87. I develop the ideas of this section and the next in another essay, "Freedoms and limitations: C. S. Lewis and Francis Schaeffer as a tag team," slated to appear in a volume honouring Professor Jerram Barrs.

88. Lewis, *Reflections on the Psalms* (London: Geoffrey Bles, 1958), ch. 11, 109.

89. Ibid.; see also Lewis, *Miracles*, ch. 4, 33.

90. Frederic Seebohm, *The Oxford Reformers: John Colet, Erasmus, and Thomas More* (London: Longmans, Green, and Co., 1869), 50–51. It is possible that Lewis mistakenly attributed this to Jerome, since he is well-known to have quoted often from memory (see Walter Hooper's "Preface" to *Selected Literary Essays*, xvii). But it is also possible that Colet did in fact derive this view from Jerome, as he did so many others. Credit for tracking down this reference goes to the blog post and comments at alltheblognamesarealreadytaken.wordpress.com/2008/05/29/cs-lewis-and-st-jerome/ (accessed May 3, 2012).

91. Lewis, *Reflections on the Psalms*, ch. 11, 110.

92. Ibid., 110–111.

93. Ibid., 115.

94. This point follows from Lewis's argument in his *Miracles*, as discussed above. See also his *Problem of Pain*, ch. 5.

95. See the discussion in Alexander Heidel, *The Babylonian Genesis* (Chicago: University of Chicago Press, 1951), which certainly fits Lewis's time frame.

96. For discussion and documentation, see my *Did Adam and Eve Really Exist? Who They Were and Why You Should Care* (Wheaton, IL: Crossway, 2011), Appendix 1.

97. I have dealt with these issues at more length in *Did Adam and Eve Really Exist*, §2b.

98. Francis A. Schaeffer, *No Final Conflict* (London: Hodder and Stoughton / Downers Grove, IL: Inter Varsity Press, 1975), ch. 3. This appears also in Francis A. Schaeffer, *The Complete Works of Francis A. Schaeffer* (Westchester, IL: Crossway, 1982), vol. 2.

99. Lewis, *Problem of Pain*, ch. 5.

100. Ibid., 70.

101. Ibid., 77–80.

102. Lewis, *Miracles*, ch. 5, 32.

103. See discussion in my *Did Adam and Eve Really Exist*, §5d; I propose my own "freedoms and limitations" in §5c. In my discussion of Lewis, I draw attention to his bafflement over the solidarity concept that underlies the Biblical picture. Hence it fascinated me to discover after my book's publication that Lewis continued to wish for greater light on this topic: In a letter written in 1951, more than 10 years after *The Problem of Pain*, he wishes he had known more about how the notion of "members of one another" works (letter, September 12, 1951, cited in Purtill, *C. S. Lewis' Case for the Christian Faith*, 56). This idea is part of what New Testament scholars now call "interchange," and it is part of the larger question of corporate solidarity among the people of God.

104. Lewis, "The Funeral of a Great Myth" (ca. 1945), *Christian Reflections*, 86. Lewis may very well be reflecting a comment by G. K. Chesterton, whom he admired: "No philosopher denies that a mystery still attaches to the two great transitions: the origin of the universe itself and the origin of the principle of life itself. Most philosophers have the enlightenment to add that a third mystery attaches to the origin of man himself. In

other words, a third bridge was built across a third abyss of the unthinkable when there came into the world what we call reason and what we call will." Chesterton, *The Everlasting Man* (Garden City: Doubleday, 1955 [1925]), 27.

105. J. B. S. Haldane, whom I have mentioned already, certainly did want to claim that evolutionary biology ruled out any possibility of conventional theistic religion—Christian, Jewish, or Islamic—being true. He states this plainly in his essays "Darwinism Today," "When I am Dead," and "Science and Theology as Art Forms" in *Possible Worlds*; see further Gordon McOuat and Mary P. Winsor, "J. B. S. Haldane's Darwinism in Its Religious Context," *British Journal for the History of Science*, 28, no. 2 (1995): 227–231.

106. Watson, "Adaptation," 88.

107. "NSTA Position Statement: The Teaching of Evolution."

108. "NABT Position Statement on Teaching Evolution."

109. Lewis, "Man or Rabbit?" (1946), *God in the Dock*, 108.

110. Lewis, *The Four Loves* (London: Geoffrey Bles, 1960), ch. 4, 70, 84. In the same chapter Lewis observes, "To those—and they are now the majority—who see human life merely as a development and complication of animal life all forms of behaviour which cannot produce certificates of an animal origin and of survival value are suspect." *Four Loves*, ch. 4, 71.

111. Susan Milius, "Beast buddies," *Science News* 164, no. 18 (November 1, 2003): 282–284.

112. Lewis, "Is Theology Poetry" (1944), *Weight of Glory*, 103; cf. "Funeral of a Great Myth" (ca. 1945), *Christian Reflections*, 89. See also "Religion without Dogma?" (1946), *God in the Dock*, 135, where Lewis notes that there are different levels of openness to religion in the different scientific disciplines: "It is as their subject matter comes nearer to man himself that their anti-religious bias hardens."

113. Information from the university's web page on Professor Cashmore, accessed May 2, 2012, http://www.bio.upenn.edu/faculty/cashmore/index.html.

114. Anthony Cashmore, "The Lucretian swerve: The biological basis of human behavior and the criminal justice system," *Proceedings of the National Academy of Sciences*, January 12, 2010 (PNAS Early Edition), accessed May 3, 2012, www.pnas.org/cgi/doi/10.1073/pnas.0915161107.

115. Lewis, *Miracles*, ch. 6, 37.

116. This last point comes from Stephen R. L. Clark, *From Athens to Jerusalem: The Love of Wisdom and the Love of God* (Oxford: Clarendon Press, 1984), 96–97; see further at 28–30 for a position similar to Lewis's. Along the same lines as Cashmore is Tamler Sommers and Alex Rosenberg, "Darwin's Nihilistic Idea: Evolution and the Meaninglessness of Life," *Biology and Philosophy* 18 (2003), 653–68. They claim that Darwin's naturalism and materialism undermine any notion of teleology for human life, and thus any notion of ethical normativity. Again, these authors are nevertheless assuming an ethics of belief, in the sense that they aim to persuade. Indeed, they get explicitly moralistic when they speak of "every naturalist's obligation to be a Darwinian about how ethical beliefs arose in the first place" (667), and later conclude, "If it is the right conclusion then we must respond to Dennett's final question… with a simple 'no'" (668). I should recommend a dose of Lewis's essays in *Christian Reflections* as a helpful antidote.

117. Lewis, *Miracles*, ch. 5, 37.

118. Lewis, *The Lion, the Witch and the Wardrobe* (New York: HarperCollins, 1950), ch. 5, 48.
119. I borrowed this list of troubles from Purtill, *C. S. Lewis' Case*, 25.
120. Lewis, *Surprised by Joy*, ch. 7, 95.
121. Farrer, "The Christian Apologist," 26.
122. Lewis, "Christian Apologetics" (1945), 103.

ORIGINS

This revolution [the adoption of an evolutionary model in biology] was certainly not brought about by the discovery of new facts.

—C. S. Lewis, The Discarded Image

6

DARWIN IN THE DOCK

C. S. LEWIS'S CRITIQUE OF EVOLUTION AND EVOLUTIONISM

John G. West

IN THE SULTRY SUMMER OF 1925, A BATTLE ROYAL RAGED IN DAYTON, Tennessee as political titans Clarence Darrow and William Jennings Bryan clashed in the trial of high school teacher John T. Scopes, an event immortalized—albeit inaccurately—in the play and film *Inherit the Wind*.[1] The Scopes trial cast a long shadow. Even today it continues to be used to caricature anyone skeptical of Darwin's theory of evolution as an unsophisticated dunce.

Few people would ever accuse C. S. Lewis of being either unsophisticated or a dunce. Yet during the summer of the Scopes trial in America, a twenty-something C. S. Lewis in England was expressing his own doubts about Darwin. A veteran of the front lines of World War I, Lewis had just been elected a Fellow of Magdalen College at Oxford. Previously a tutor in philosophy, his new position was in English literature.[2]

A few weeks after the Scopes trial concluded, Lewis wrote his father about his change in academic fields. He commented that although he was glad of the change, he was grateful for "something of value" that he learned from philosophy: "It will be a comfort to me all my life to know that the scientist and the materialist have not the last word: that Darwin and Spencer undermining ancestral beliefs stand themselves on a foundation of sand; of gigantic assumptions and irreconcilable contradictions an inch below the surface."³ Still an atheist, Lewis already had gnawing doubts about Darwinism.

Lewis's early skepticism of Darwinism makes it all the more astonishing that he has been honored as a veritable patron saint in recent years by some contemporary proponents of theistic evolution. In the best-selling book *The Language of God* (2006), for example, biologist Francis Collins highlighted the role Lewis's writings played in his own conversion to Christianity as well as invoking Lewis to defend the idea that Christians should accept the animal ancestry of humans.⁴ In a 2010 article in the journal *Perspectives on Science and Christian Faith*, philosopher Michael Peterson of Asbury University went considerably further. According to Peterson, Lewis not only embraced "both cosmic and biological evolution as highly confirmed scientific theories," but he would have rejected out-of-hand arguments offered by modern proponents of intelligent design.⁵ In 2011, Peterson's article on Lewis and evolution was serialized online by the pro-theistic-evolution group BioLogos.⁶

On one level, appeals to Lewis by proponents of theistic evolution are easy to understand. Despite more than a century of boosterism by Darwin's defenders, many traditional Christians remain deeply skeptical of efforts to mix God and Darwinian evolution. Indeed, according to a nationwide survey of Protestant clergy released in 2012, an overwhelming 72% disagreed with the position that "God used evolution to create people."⁷

The skepticism of theistic evolution by many Christians is fueled by leading theistic evolutionists who challenge Biblical authority, dismiss the historicity of the Fall, and even deny that God specifically directs the evolutionary process. According to Biblical studies professor Kenton Sparks (writing for the BioLogos website), "If Jesus as a finite human being erred from time to time, there is no reason at all to suppose that Moses, Paul, John wrote Scripture without error."⁸ According to Karl Giberson (a co-founder of BioLogos with Francis Collins), human beings were selfish and sinful from the very start, so there was no actual "Fall."⁹ And according to biologist Kenneth Miller, author of *Finding*

Darwin's God (a seminal text for modern American theistic evolutionists), God himself did not know whether the "undirected" process of evolution would produce human beings or something radically different, say, "a mollusk with exceptional mental capabilities."[10]

In contrast to such heterodoxy, C. S. Lewis is a beloved icon of mainstream, historic Christianity. He provides a "Good Housekeeping" seal of approval for many Christians. If Lewis can be enlisted as a supporter of theistic evolution and a critic of intelligent design, perhaps theistic evolution will be able to broaden its base of support among traditional Christians.

There is little doubt that Lewis was interested in the topic of evolution. He discussed it repeatedly in his books and essays, although circumspectly. He wrote about it in his private letters. And his personal library contained more than a dozen books and pamphlets focused on evolution, some of which were marked up with extensive underlining and annotations, including his personal copy of Charles Darwin's *Autobiography*.[11]

Although Lewis was interested in evolution, he also understood its cultural dominance, which helps explain his cageyness in publicly communicating about the topic. He observed to one correspondent that "Evolution etc" is the "assumed background" of modern thought.[12] He declined to write a preface to an anti-evolution book by someone else because he feared the repercussions for his public role as a Christian apologist: "Many who have been or are being moved towards Christianity by my books wd. be deterred by finding that I was connected with anti-Darwinism."[13]

If Lewis was cautious about how much he criticized Darwinian evolution in public, he was equally careful to distance himself from evolution's uncritical boosters. Michael Peterson quotes Lewis in *Mere Christianity* as flatly affirming that "Everyone now knows... that man has evolved from lower types of life," as if Lewis thought no reasonable person could disagree.[14] But this is a case of putting words in Lewis's mouth through creative editing.

Here is the unedited version of what Lewis actually wrote in *Mere Christianity* with the words Peterson deleted in bold: "Everyone now knows **about Evolution (though, of course, some educated people disbelieve it):** everyone has been told that man has evolved from lower types of life."[15] Reading the sentence in its entirety, one can see that far from asserting that "Evolution" is something that "everyone now knows," Lewis was merely stating that "everyone now knows *about* Evolution," and "everyone has been *told*" certain things about it. This was a description of popular sentiment, not a statement about whether evolution is true or false. Lest someone misunderstand Lewis as endorsing the idea that no reasonable person can doubt evolution, Lewis added the caveat, "of course, some educated people disbelieve it."

Lewis's reservations about evolution in this passage were quite intentional, as they were inserted by him after the original radio talk that formed the basis of this section of *Mere Christianity*.[16] Nevertheless, Lewis's overall point in this passage had nothing to do with biological evolution at all. He invoked evolution as an analogy to help people better understand the Christian doctrine of sanctification. He cited evolution because he thought it was an illustration "a modern man can understand."

So what *were* Lewis's real views about evolution? To answer that question fairly, we first need to untangle the distinct ways in which Lewis employed the term.

One of the most challenging things about discussing "evolution" today is that the term is so elastic, covering everything from mere "change over time" to the development of all living things from one-celled organisms to man through an unguided process of natural selection acting on random variations. Evolution has so many different meanings, in fact, that if one doesn't pay close attention, a conversation on the topic will quickly devolve into people talking past one another.[17]

Lewis addressed at least three different kinds of evolution in his writings: (1) evolution as a theory of common descent; (2) evolution as a theory of unguided natural selection acting on random variations (a.k.a.

Darwinism); and (3) evolution as a cosmic philosophy (a.k.a. "evolutionism").

As we shall see, Lewis did not object in principle to evolution in the first sense (common descent), although he sharply limited its application in a way that mainstream proponents of evolution would find unacceptable. The case for Lewis as a supporter of evolution in the second sense (Darwinism) is almost non-existent. Lewis was a thoroughgoing skeptic of the creative power of unguided natural selection. As for evolution in the third sense—evolutionism—Lewis respected the poetry and grandeur of what he sometimes called the "myth" of evolution, but he certainly regarded it as untrue.

Lewis's Limited Acceptance of Common Descent

Common descent is the claim that all organisms currently living have descended from one or a few original ancestors through a process Darwin called "descent with modification." According to this idea, not only humans and apes share an ancestor, but so do humans, clams, and fungi. Common descent is a hallowed dogma among today's evolution proponents, held with quasi-religious fervor.

Lewis clearly believed that Christians can accept evolution as common descent without doing violence to their faith. This is what Lewis was getting at when he wrote to evolution critic Bernard Acworth, "I believe that Christianity can still be believed, even if Evolution is true."[18] In Lewis's view, whether God used common descent to create the first human beings was irrelevant to the truth of Christianity. As he wrote to one correspondent late in his life, "I don't mind whether God made man out of earth or whether 'earth' merely means 'previous millennia of ancestral organisms.' If the fossils make it probable that man's physical ancestors 'evolved,' no matter."[19]

In *The Problem of Pain* (1940), Lewis even offers a possible evolutionary account of the development of human beings, although he makes clear he is offering speculation, not history: "[I]f it is legitimate to guess," he writes, "I offer the following picture—a 'myth' in the Socratic sense," which he defines as "a not unlikely tale," or "an account of what *may have been* the historical fact" (emphasis in the original). Lewis then suggests that "[f]or long centuries God perfected the animal form which was to become the vehicle of humanity and the image of Himself… The creature may have existed for ages… before it became man."[20] Elsewhere, Lewis seemed smitten by the idea of embryonic recapitulation, the discredited evolutionary idea that human beings replay the history of their evolution from lower animals in their womb. And in a letter to his friend Anglican Nun Sister Penelope in 1952, he mentioned his previous speculation that the first human being was descended from "two anthropoids."[21]

Nevertheless, Lewis did not exactly go out of his way to champion the animal ancestry of humans. When pressed on the subject by evolution critic Bernard Acworth in the 1940s, Lewis backpedaled, replying that his "belief that Men in general have immortal & rational souls does not oblige or qualify me to hold a theory of their pre-human organic history—if they have one."[22] A few years later, Lewis relished the exposure of "Piltdown Man" as a hoax. Originally touted as evidence for the long-sought "missing link" between apes and humans, the Piltdown Man's skull was discovered in the 1950s to be a fake forged from the skull of a modern human, the jawbone of an orangutan, and the teeth of a chimpanzee.[23] Lewis wrote to Bernard Acworth that although he didn't think the scandal should be exploited, "I can't help sharing a sort of glee with you about the explosion of poor old Piltdown… one inevitably feels what fun it wd. be if this were only the beginning of a landslide."[24] He wrote another correspondent, "The detection of the Piltdown forgery was fun, wasn't it?"[25] Interestingly, four years before the definitive exposure of Piltdown as a fraud, Lewis had already published a poem that labeled the fossil the "fake from Piltdown."[26] His final Narnian story, meanwhile, completed a few months after the Piltdown scandal hit the headlines, features as the villain an ape who insists he is really a human being—perhaps Lewis's whimsical commentary on "poor old Piltdown."[27]

Whatever Lewis's final position on the animal ancestry of the human race, it would be wrong to conclude that his acceptance of some kind of human evolution placed him in the camp of mainstream evolutionary biology, or even of mainstream theistic evolution. In fact, Lewis insisted on three huge exceptions to evolutionary explanations of humanity that placed him well outside evolutionary orthodoxy, both then and now.

An Historic Fall

LEWIS'S FIRST exception to human evolution was his insistence on an actual Fall of Man from an original state of innocence. In Christian theology, God originally created human beings morally innocent. These first humans then freely rejected God's will for them, resulting in a Fall from innocence and harmony into the sinful condition of the human race as we currently find it. According to historic Christian teaching, not only human beings, but the entire creation was tainted by man's initial act of wrongdoing. It was to reverse the impact of the Fall that God became incarnate to save us from our sins. Thus, the Fall provides the necessary "back story" for Jesus Christ and his death on the cross.

Leading theistic evolutionists no less than secular evolutionists insist that an historic Fall is incompatible with mainstream evolutionary theory. In the words of Anglican Bishop John Shelby Spong, "Darwin... destroyed the primary myth by which we had told the Jesus story for centuries. That myth suggested that there was a finished creation from which we human beings had fallen into sin, and therefore needed a rescuing divine presence to lift us back to what God had originally created us to be. But Charles Darwin says that there was no perfect creation." Thus, "there was no perfect human life which then corrupted itself and fell into sin... And so the story of Jesus who comes to rescue us from the fall becomes a nonsensical story."[28]

Spong is well known for being a theological liberal, but similar views are gaining prevalence among evangelical Christian proponents of evolution. Karl Giberson, a co-founder with Francis Collins of the pro-theistic-evolution group BioLogos, likewise repudiates the traditional teaching that "sin originates in a free act of the first humans" and that "God gave humans free will and they used it to contaminate the entire creation."[29] In his book *Saving Darwin*, Giberson has a section titled "Dissolving the Fall" where he essentially argues that since human beings were created through Darwinian evolution, they were never morally good. Instead, they were sinful from the start because the evolutionary process is based on selfishness: "Selfishness... drives the evolutionary process. Unselfish creatures died, and their unselfish genes perished with them. Selfish creatures, who attended to their own needs for food, power, and sex, flourished and passed on these genes to their offspring. After many generations selfishness was so fully programmed in our genomes that it was a significant part of what we now call human nature."[30] Francis Collins wrote an enthusiastic foreword to Giberson's book.

Lewis was well aware of the problems posed by mainstream evolutionary theory for the Christian concept of the Fall. His personal library included a copiously underlined copy of *The Unveiling of the Fall* (1923) by the Rev. C. W. Formby, which forcefully laid out the incompatibility of evolution and the traditional Christian belief that human beings and the world were originally created morally good.[31] Lewis's underlining of the book included the following passage outlining the sinful tendency of the evolutionary process as a whole: "Obviously this entire organic process, if not actually sin-producing, is, according to its natural self-centred principles, certainly conducive to sin, and has never ceased to manifest signs of this fact."[32] Accordingly, the evolutionary view as applied to man "places him before us already burdened with an inherently self-centred nature, dominated by those instinctive structures of animalism whose overpowering bias toward evil even to-day, after centuries of civilisation and restraint, is still sometimes irresistible. Thus, this theory puts man before us in a practically fallen condition from the start."[33] Rev. Formby thought this view was theologically untenable because it forced us to adopt "the impossible belief that both sin and suffering came into existence as a practically unavoidable outcome of God's direct action."[34]

Despite this apparent incompatibility of the evolutionary account with good theology, Formby was loath to disown either the Fall or evolution. Instead, he pulled the proverbial rabbit out of the hat and proposed a *pre-organic* fall of human beings.[35] That is, in his view the first human beings previously existed as spiritual beings and fell from grace *before* they became embodied. The pain and suffering brought about by evolution was therefore excusable because humans as well as animals were already fallen, and in a fallen state pain and suffering are used by God to bring fallen creatures back to him. Lewis refrained from adopting Formby's heterodox explanation, although he did suggest that the fall of Satan and his fellow angels had something to do with pain and suffering among the lower animals.[36] But regarding humans, Lewis insisted there was a real fall inside human history. He further made clear that this belief was non-negotiable in his view for orthodox Christians.

Noting that "[i]t is not yet obvious to me that all theories of evolution do contradict" the Fall, Lewis was emphatic that any evolutionary theory that does deny a real Fall is unacceptable: "I believe that Man has fallen from the state of innocence in which he was created: I therefore disbelieve in any theory wh. contradicts this."[37] Accordingly, Lewis was careful in *The Problem of Pain* to preserve an historical Fall as part of his hypothetical account of human evolution. Indeed, he titled the chapter in which his evolutionary account appears "The Fall of Man," and at the end of that chapter he declared that "the thesis of this chapter is simply that man, as a species, spoiled himself."[38] Following traditional Christian teaching, Lewis emphasized that man prior to the Fall had unimpeded fellowship with God. "God came first in his love and in his thought, and that without painful effort. In perfect cyclic movement, being, power and joy descended from God to man in the form of gift and returned from man to God in the form of obedient love and ecstatic adoration."[39] Lewis acknowledged that pre-Fall man might look crude when "[j]udged by his artefacts, or perhaps even by his language," and he did "not doubt that if the Paradisal man could now appear among us, we should regard him as an utter savage." Yet Lewis added that upon taking a second look, "the holiest among us... would fall at his feet."[40]

Lewis's account of human life before the Fall is worth close attention. He suggested that man in his original state lived in complete harmony with himself and his surroundings. Before the Fall, man's judgment exercised complete command of his appetites. Sleep was "not the stupor which we undergo, but willed and conscious repose." Lifespans were under man's control: "Since the processes of decay and repair in his tissues were similarly conscious and obedient, it may not be fanciful to suppose that the length of his life was largely at his own discretion." And man lived in harmony with the animals: "Wholly commanding himself, he commanded all lower lives with which he came into contact. Even now we meet rare individuals who have a mysterious power of taming beasts. This power the Paradisal man enjoyed in eminence. The old picture of the brutes sporting before Adam and fawning upon him may not be wholly symbolical."[41]

Lewis's description of human life before the Fall sounds very much like the literal "Eden" described by historic Christian teaching. Lewis embraced the essential reality of Eden, as did his close friend J. R. R. Tolkien, whose views on the matter were influenced by Lewis. According to Tolkien, Eden did not have an "historicity of the same kind as the N[ew] T[estament]," but it nevertheless really existed. "Genesis is separated by we do not know how many sad exiled generations from the Fall, but certainly there was an Eden on this very unhappy earth. We all long for it, and we are constantly glimpsing it: our whole nature at its best and least corrupted, its gentlest and most humane, is still soaked with the sense of 'exile.'"[42] In Tolkien's view, every expression of horror at some evil, as well as every idyllic memory of our home life, is "derived from Eden. As far as we can go back the nobler part of the human mind is filled with the thoughts of... peace and goodwill, and with the thought of its *loss*."[43] Tolkien bristled at how scientists had successfully brow-beaten Christians into disowning their belief in the reality of Eden: "As for Eden, I think most Christians... have been rather bustled and hustled now for some generations by the self-styled scientists [who have]... tucked Genesis into a lumber-room of their mind as not very fashionable furniture, a bit ashamed to have it about the house, don't you know, when the bright clever young people called." But Tolkien added that he now no longer felt "either ashamed or dubious" about his belief in "the Eden 'myth.'"[44] He attributed his change of heart partly to his interactions with Lewis.

Lewis had little patience for those evolutionists (theistic or otherwise) who asserted that modern science made it impossible to believe in man's original Paradisal state and subsequent fall. At the heart of their assertions, in Lewis's view, was what he called "the idolatry of artefacts"[45]—the assumption that we can discern the morality or intelligence of ancient peoples from their material products. Lewis pointed out that pottery shards or spear-heads might expose the primitive state of a prehistoric people's technology, but they do nothing to reveal the state of the people's morality or even their native intelligence. Such archeological discoveries do not tell us whether prehistoric peoples were kind, or courageous, or noble, or just. Nor do they tell us about their capacity for poetry or song, let alone technological innovation. "What is learned by trial and error must begin by being crude, whatever the character of the beginner," wrote Lewis. So "[t]he very same pot" that "would prove its maker a dunce if it came after millenniums of pot-making" also "would prove its maker a genius if it were the first pot ever made in the world." Consequently, genuine "[s]cience... has nothing to say either for or against the doctrine of the Fall."[46]

If Lewis dismissed claims that science refuted the Fall, he was equally skeptical of efforts to reinterpret the Fall to make it part of evolutionary history. In the standard evolutionary picture (popularized by Darwin himself in *The Descent of Man*), human beings started out as brutes and only gained morality and religion after a long struggle for survival.[47] Given this view of the development of human beings, it is hardly surprising that some theistic evolutionists have concluded that if there was a "Fall" in evolutionary history, it must have been a "fall upward" into greater maturity and responsibility of the sort advocated by liberal theologians since Hegel and Kant. For example, contemporary Christian thinker Brian McLaren argues that the Fall is best understood not as a fall from a higher state of innocence and goodness, but as a "compassionate coming of age story" that represents "the first stage of ascent as human beings progress from the life of hunter-gatherers to the life of agriculturalists and beyond."[48] McLaren does acknowledge that the ascent of man is marked by struggles with sin. But he seems to believe that human wrongdoing is a natural part of God's plan to bring about human maturity. Lewis spent much of his novel *Perelandra* (1943) critiquing this kind of thinking, arguing that God intended for human beings to progress to self-knowledge and maturity by obedience, not rebellion.[49] Four years later in his book *Miracles* (1947), Lewis ridiculed those who "say that the story of the Fall in *Genesis* is not literal; and then go on to say (I have heard them myself) that it was really a fall upwards—which is like saying that because 'My heart is broken' contains a metaphor, it therefore means 'I feel very cheerful'. This mode of interpretation I regard, frankly, as nonsense."[50]

Lewis continued to defend the reality of the Fall to those who corresponded with him. "I'm not a Fundamentalist in the strict sense... But I often agree with the Fundamentalists about particular passages where literal truth is rejected by many moderns," Lewis wrote to a correspondent in 1955. Lewis went on to reaffirm his belief in "the Fall" and echo his argument from *The Problem of Pain* by stating that "I don't see what the findings of the scientists can say either for or against it. You can't see from looking at skulls or flint implements whether man fell or not." He then referred his correspondent to *The Problem of Pain* as well as "G. K. Chesterton['s] *The Everlasting Man* which is excellent on this point."[51] To another correspondent who questioned the grounds of Lewis's belief that the earliest humans lived unfallen in a paradise-like state, Lewis replied: "[Y]ou *do* know very well what grounds I have for assuming the existence of *Paradisal Man*—namely that it is part of orthodox Christianity."[52]

A Literal Adam

LEWIS NOT only believed in an historic Fall; he also embraced the literal existence of Adam and Eve, which was another important exception to his acquiescence to human evolution. Lewis's acceptance of an historical Adam and Eve is widely unrecognized today. Popular Christian pastor Tim Keller, for example, writes that "C. S. Lewis... did not believe in a literal Adam and Eve."[53] Keller is misinformed, at least when it comes to Lewis's beliefs after he became a Christian. While Lewis was still a young atheist in the 1920s, he certainly disbelieved in Adam and Eve, although he was simultaneously skeptical of orthodox Darwinism.[54] By the 1940s, however, he was publicly noncommittal, writing in *The Problem of Pain* that "we do not know how many of these [unfallen] creatures God made, nor how long they continued in the Paradisal state."[55] In private, he was not so reticent. In a discussion at his home attended by Oxford colleague Helen Gardner, Lewis stated that the person from history he would most like to meet in heaven was Adam. When Gardner protested that "if there really were, historically, someone whom we could name as 'the first man', he would be a Neanderthal ape-like figure, whose conversation she could not conceive of finding interesting," Lewis is said to have responded with disdain: "I see we have a Darwinian in our midst."[56]

It is worth noting that throughout Lewis's imaginative works, Adam and Eve are typically treated as real figures from history, not as allegories or myths, even when the characters in Lewis's stories are seeking to explain truths about the "real" world. In the Narnian Chronicles, human beings are repeatedly referred to as "Sons of Adam" and "Daughters of Eve," and during Lewis's telling of a temptation story on another planet in *Perelandra*, the hero repeatedly affirms the teachings of traditional theology to the planet's equivalent of Eve, including a traditional account of Adam and Eve: "Long ago, when our world began, there was only one man and one woman in it, as you and the King are in this. And there once before he [the Tempter] stood, as he stands now, talking to the woman... And she listened, and did the thing Maleldil [God] had forbidden her to do. But no joy and splendour came of it."[57]

Additionally, Lewis treated Adam as a real person in history in his private correspondence. To his friend St. Giovanni Calabria, an Italian priest, he wrote about the "necessary doctrine that we are most closely joined together alike with the sinner Adam and with the Just One, Jesus,"[58] while to another correspondent he described his novel *Perelandra* as the working out of the "supposition" that what happened to Adam and Eve on earth could happen to another first couple elsewhere: "Suppose, even now, in some other planet there were a first couple undergoing the same [temptation] that Adam and Eve underwent here, but successfully."[59]

A Mindless Process Could Not Produce Man

LEWIS'S FINAL exception to human evolution was his insistence that the development of human beings required far more than a mindless material process. In his own words, his speculations about human evolution "had pictured Adam as being, physically, the son of two anthropoids, on whom, after birth, God worked the miracle which made him Man."[60] In Lewis's view, Darwinian evolution might possibly explain man's physical form; but it could not explain man's mind, his morality, or his eternal soul. That is because the driving force of modern Darwinism was supposed to be the mindless mechanism of natural selection acting on random variation, and Lewis was deeply skeptical about what such a mindless mechanism could actually achieve.

Lewis's Doubts about the Creative Power of Natural Selection

LEWIS KNEW THAT THE TRULY momentous feature of modern evolutionary theory is not its proposal that life has a long history, nor even its claim that humans and apes share a common ancestor. No, the truly radical part of modern evolutionary theory is its insistence that life is the product of an unguided process. The claim that evolution is the product of chance and necessity is not just the product of the fevered imaginations of muscular "New Atheists" like biologist Richard Dawkins. It forms the very core of orthodox Darwinian theory, which claims that the primary driver of evolution is an unguided process of natural selection (or "survival of the fittest") operating on random variations in nature (random mutations, according to modern evolutionists).

Darwin himself repeatedly made clear that evolution by natural selection neither required nor involved intelligent guidance. Indeed, according to Darwin, his theory of natural selection provided a definitive refutation of the idea that the features of the natural world reflected a preconceived design:

> The old argument of design in nature, as given by Paley, which formerly seemed to me so conclusive, fails, now that the law of natural selection has been discovered. We can no longer argue that, for instance, the beautiful hinge of a bivalve shell must have been made by an intelligent being, like the hinge of a door by man. There seems to be no more design in the variability of organic beings and in the action of natural selection, than in the course which the wind blows.[61]

If natural selection was unguided in Darwin's view, so too were the variations in nature on which selection acted. Objecting to those who claimed that beneficial variations in nature might be the result of intelligent design, Darwin declared:

> no shadow of reason can be assigned for the belief that variations… which have been the groundwork through natural selection of the formation of the most perfectly adapted animals in the world, man included, were intentionally and specially guided. However much we may wish it, we can hardly follow Professor Asa Gray in his belief "that variation has been led along certain beneficial lines," like a stream "along definite and useful lines of irrigation."[62]

The dominant view of evolution today in the scientific community remains essentially Darwinian. In the words of 38 Nobel laureates who issued a statement defending Darwin's theory in 2005, evolution is "the result of an *unguided, unplanned* process of random variation and natural selection."[63]

One certainly can conceive of a theory of guided evolution, but mainstream Darwinian theory is not it. *Darwinian* evolution by definition is an unguided process that brings forth new things through a combination of chance and necessity. But can such a fundamentally mindless and undirected process create the exquisite form and function seen throughout the natural world? Lewis was skeptical.

Lewis did affirm that "[w]ith Darwinianism as a theorem in Biology I do not think a Christian need have any quarrel."[64] But for Lewis "Darwinianism as a theorem in Biology" was a pretty modest affair. Contra leading evolutionists, Lewis thought the "purely biological theorem... makes no cosmic statements, no metaphysical statements, no eschatological statements." Nor can Darwinism as a scientific theory explain many of the most important aspects of biology itself: "It does not in itself explain the origin of organic life, nor of the variations, nor does it discuss the origin and validity of reason." So what *can* the Darwinian mechanism explain according to Lewis? "Granted that we now have minds we can trust, granted that organic life came to exist, it tries to explain, say, how a species that once had wings came to lose them. It explains this by the negative effect of environment operating on small variations."[65] In other words, according to Lewis, Darwin's theory explains how a species can change over time by losing functional features it already has. Suffice to say, this is not the key thing the modern biological theory of evolution purports to explain. Noticeably absent from Lewis's description is any confidence that Darwin's unguided mechanism can account for the formation of fundamentally new forms and features in biology. Natural selection can knock out a wing, but can it build a wing in the first place? Lewis didn't seem to think so.

A further indication of just how skeptical Lewis was about the creative power of natural selection appears in a talk he delivered to the Oxford University Socratic Club in 1944. There Lewis stated that "[t]he Bergsonian critique of orthodox Darwinism is not easy to answer."[66] Lewis was referring to Henri Bergson (1859–1941), a French natural philosopher and Nobel laureate who offered a decidedly non-Darwinian account of evolution in his book *L'Evolution Creatice* (*Creative Evolution*).[67]

Lewis first read Bergson during World War I while recovering from shrapnel wounds from the front lines, and the experience on Lewis was profound. In his autobiography *Surprised by Joy*, Lewis said that Bergson "had a revolutionary effect on my emotional outlook... From him I first learned to relish energy, fertility, and urgency; the resource, the triumphs, and even the insolence, of things that grow." Lewis also was grateful to Bergson for making him "capable of appreciating artists who would, I believe, have meant nothing to me before; all the resonant, dogmatic, flaming, unanswerable people like Beethoven, Titian (in his mythological pictures), Goethe, Dunbar, Pindar, Christopher Wren, and the more exultant Psalms."[68] Lewis continued to re-read Bergson in the years that followed as he continued his studies at Oxford. During the summer of 1920, he wrote a friend that he was "reading Bergson now and find all sort of things plain sailing which were baffling a year ago."[69] A year earlier, he wrote his father that he was living in anticipation of a visit to Oxford by Bergson, but commented wistfully that "I suppose I shall not see [him]... unless he gives a lecture."[70] The impact of Bergson on Lewis is indicated in Lewis's 1917 copy of *L'Evolution Creatice*, which is filled with careful annotations and underlining on most of its nearly 400 pages.[71]

Bergson was an unsparing critic of the creative power of Darwinian natural selection. Granting that "[t]he Darwinian idea of adaptation by automatic elimination of the unadapted is a simple and clear idea," he argued that precisely "because it attributes to the outer cause which controls evolution a merely negative influence, it has great difficulty in accounting for the progressive and, so to say, rectilinear development of complex apparatus" like the vertebrate eye.[72] Bergson stressed that Darwinism's reliance on accidental variations as the raw material for evolution made the development of highly coordinated and complex features found in biology nothing short of incredible. This was the case regardless of whether the accidental variations were slight or large.

As Bergson noted, some Darwinians insisted that the variations used by evolution were so slight that they would not hinder the survival of the organism: "For a difference which arises accidentally at one point of the visual apparatus, if it be very slight, will not hinder the functioning of the organ; and hence this first accidental variation can, in a sense, wait for complementary variations to accumulate and raise vision to a higher degree of perfection." Bergson granted the point, but then noted the problem it raised: "[W]hile the insensible variation does not hinder the functioning of the eye, neither does it help it, so long as the variations that are complementary do not occur. How, in that case, can the variation be retained by natural selection? Unwittingly one will reason as if the slight variation were a toothing stone set up by the organism and reserved for a later construction." But "[t]his hypothesis" is obviously "little conformable to the Darwinian principle," which emphasizes that natural selection acts mechanically and without foresight.[73] To get around this problem, other Darwinists claimed that evolution relied on large accidental variations that provided evolutionary leaps. "But here there arises another problem, no less formidable," wrote Bergson, "viz., how do all the parts of the visual apparatus, suddenly changed, remain so well coordinated that the eye continues to exercise its function? For the change of one part alone will make vision impossible, unless this change is absolutely infinitesimal. The parts must then all change at once, each consulting the others." Even "supposing chance to have granted this favour once, can we admit that it repeats the self-same favour in the course of the history of a species, so as to give rise, every time, all at once, to new complications marvellously regulated with reference to each other, and so related to former complications as to go further on in the same direction?"[74]

The sheer improbability of the Darwinian explanation increases exponentially once one realizes how frequently the same complex biological features are supposed to have arisen independently in different evolutionary lineages. In the words of Bergson: "What likelihood is there that, by two entirely different series of accidents being added together, two entirely different evolutions will arrive at similar results?"[75] The whole idea was incredible according to Bergson:

> An accidental variation, however minute, implies the working of a great number of small physical and chemical causes. An accumulation of accidental variations, such as would be necessary to produce a complex structure, requires therefore the concurrence of an almost infinite number of infinitesimal causes. Why should these causes, entirely accidental, recur the same, and in the same order, at different points of space and time?

Responding to his own question, Bergson replied that "[n]o one will hold that this is the case, and the Darwinian himself will probably merely maintain that identical effects may arise from different causes, that more than one road leads to the same spot." But this was fallacious reasoning: "[L]et us not be fooled by a metaphor. The place reached does not give the form of the road that leads there; while an organic structure is just the accumulation of those small differences which evolution has had to go through in order to achieve it." Hence, "[t]he struggle for life and natural selection can be of no use to us in solving this part of the problem, for we are not concerned here with what has perished, we have to do only with what has survived."[76]

From the extensive annotations Lewis made in his personal copy of *L'Evolution Creatice*, it is clear that he understood and appreciated Bergson's critique of natural selection. Lewis aptly summarized the Darwinian mechanism of adaptation according to Bergson as the "[e]limination of the unfit" and noted that it "plainly cannot account for complicated similarities on divergent lines of evolution."[77] Lewis also noted Bergson's view that "pure Darwinism has to lean on a marvellous series of accidents" and how Darwinists try to "escape" this truth "by a bad metaphor."[78] Lewis paid particular attention to Bergson's critique of Darwinian accounts of eye evolution in mollusks and vertebrates, concluding that "[n]atural selection… fails to explain these Eyes."[79]

Bergson's critique of natural selection likely paved the way for Lewis's doubts about Darwin, and may help explain Lewis's comment to his father in 1925 that "Darwin and Spencer... stand themselves on a foundation of sand."[80] But Lewis's skepticism toward natural selection was fueled by more than Bergson.

The ultimate challenge to Darwinian natural selection in Lewis's view was man himself. How could such a blind material process produce man's unique capabilities of reason and conscience? Lewis, of course, was far from the first intellectual to doubt Darwinism's ability to explain man. Alfred Russel Wallace, co-founder with Darwin of the modern theory of evolution itself, raised the same doubts, as did Roman Catholic zoologist St. George Jackson Mivart, whose best-selling book *The Genesis of Species* gave Darwin fits. To rebut the naysayers, Darwin responded in 1871 with two volumes and nearly 900 pages of prose in his treatise *The Descent of Man*, which forcefully argued that unguided natural selection could produce man's mental and moral faculties perfectly well, thank you.

Lewis thought otherwise, and he was tutored in his doubts by a book from of one of his favorite authors, G. K. Chesterton. The book was Chesterton's *The Everlasting Man* (1922), which Lewis read for the first time in the mid-1920s. Near the end of his life, Lewis placed *The Everlasting Man* on a list of ten books that "did [the] most to shape" his "vocational attitude and... philosophy of life." In chapter 2 of *The Everlasting Man* ("Professors and Prehistoric Men"), Chesterton skewered the pretensions of anthropologists who spun detailed theories about the culture and capabilities of primitive man based on a few flints and bones, likely inspiring Lewis's discussion of "the idolatry of artefacts" in *The Problem of Pain*. But Chesterton also provides in his book a full-throttled argument as to why Darwinism cannot explain the higher capabilities of man. In Chesterton's words, "Man is not merely an evolution but rather a revolution" whose rational faculties far outstrip those seen in the other animals. Chesterton acknowledged the possibility that man's "body may have been evolved from the brutes," but he insisted that "we know nothing of any such transition that throws the smallest light upon his soul as it has shown itself in history."[81] Again: "There may be a broken trail of stones and bones faintly suggesting the development of the human body. There is nothing even faintly suggesting such a development of the human mind."[82]

Chesterton's book prepared the ground for Lewis's own eventual critique of natural selection with regard to man—as did a lesser-known volume, *Theism and Humanism* (1915) by Sir Arthur Balfour. Balfour, best remembered today as the British Prime Minister who issued the Balfour Declaration, adapted *Theism and Humanism* from the Gifford Lectures he had presented at the University of Glasgow in 1914. Balfour's goal was to show his audience "that if we would maintain the value of our highest beliefs and emotions, we must find for them a congruous origin. Beauty must be more than accident. The source of morality must be moral. The source of knowledge must be rational." Balfour thought that once this argument "be granted, you rule out Mechanism, you rule out Naturalism, you rule out Agnosticism; and a lofty form of Theism becomes, as I think, inevitable."[83] With regard to the human mind, Balfour argued that any effort to explain mind in terms of blind material causes was self-refuting: "[A]ll creeds which refuse to see an intelligent purpose behind the unthinking powers of material nature are intrinsically incoherent. In the order of causation they base reason upon unreason. In the order of logic they involve conclusions which discredit their own premises."[84] Balfour offered a similar critique of materialistic accounts of human morality, which he thought destroyed morality by depicting it as the product of processes that are essentially non-moral. Balfour takes special aim throughout his book at Darwinian explanations of mind and morals.

It is not known when exactly Lewis first came across *Theism and Humanism*. His father Albert owned a copy of a previous Balfour book, *The Foundations of Belief* (1895), but Lewis's first known mention of *Theism and Humanism* was in a lecture in the 1940s.[85] He later listed it as one of the books that influenced his philosophy of life the most,[86] and its basic arguments are on prominent view in Lewis's *Miracles: A Preliminary Study* (1947). As Paul Ford points out, "[T]he thesis and even the language of Balfour's first Gifford lectures permeates the first five chapters of *Miracles*."[87]

The revised 1960 edition of *Miracles* is generally recognized as presenting Lewis's most mature critique of the ability of naturalism/materialism to account for man's rational faculties. What is less noticed is the challenge Lewis's book raises for Darwinian evolution in particular. Theistic evolutionists like Michael Peterson prefer to treat Lewis's argument in *Miracles* as dealing merely with generic philosophical naturalism. But the specific example of naturalism Lewis attacks at length in his book is Darwinian natural selection, not plain vanilla naturalism.

In the words of Lewis, naturalists argue that "[t]he type of mental behavior we now call rational thinking or inference must… have been 'evolved' by natural selection, by the gradual weeding out of types less fitted to survive."[88] Lewis flatly denied that such a Darwinian process could have produced human rationality: "[N]atural selection could operate only by eliminating responses that were biologically hurtful and multiplying those which tended to survival. But it is not conceivable that any improvement of responses could ever turn them into acts of insight, or even remotely tend to do so." This is because "[t]he relation between response and stimulus is utterly different from that between knowledge and the truth known."[89] Natural selection could improve our responses to stimuli from the standpoint of physical survival without ever turning them into *reasoned* responses. Following Balfour, Lewis goes on to argue that attributing the development of human reason to a non-rational process like natural selection ends up undermining our confidence in reason itself. After all, if reason is merely an unintended by-product of a fundamentally non-rational process, what grounds do we have left for regarding its conclusions as objectively true?

Lewis knew that the corrosive impact of a Darwinian account of the mind was not merely theoretical. In his personal copy of Darwin's *Autobiography*, he highlighted two passages where Darwin questioned whether the conclusions of a mind produced by a Darwinian process could in fact be trusted. In the first passage, Darwin acknowledged "the extreme difficulty or rather impossibility of conceiving this immense and wonderful universe, including man... as the result of blind chance or necessity. When thus reflecting, I feel compelled to looked to a First Cause having an intelligent mind in some degree analogous to that of man; and I deserve to be called a Theist." Darwin claimed that this conclusion "was strong in my mind about the time... when I wrote the *Origin of Species*," although "since that time... it has very gradually, with many fluctuations, become weaker." As a result, he now "must be content to remain an agnostic." Why had Darwin's confidence in the existence of a First Cause collapsed? Apparently because he realized the implications of his theory for the human mind: "But then arises the doubt—can the mind of man, which has, as I fully believe, been developed from a mind as low as that possessed by the lowest animals, be trusted when it draws such grand conclusions?"[90] Lewis placed an "x" next to this revealing admission by Darwin, and he underlined an even stronger statement by Darwin making the same point three pages later. In a passage from a letter written in 1881, Darwin expressed his inconstant belief "that the Universe is not the result of chance" and then added: "But then with me the <u>horrid doubt always arises whether the convictions of man's mind, which has been developed from the mind of the lower animals, are of any value or at all trustworthy. Would any one trust in the convictions of a monkey's mind, if there are any convictions in such a mind?</u>"[91] (underlining by Lewis)

Lewis argued that the theist need not suffer such paralyzing doubts because "[h]e is not committed to the view that reason is a comparatively recent development moulded by a process of selection which can select only the biologically useful. For him, reason—the reason of God—is older than Nature, and from it the orderliness of Nature, which alone enables us to know her, is derived." Thus, "the preliminary processes within Nature which led up to" the human mind—"if there were any"—"were designed to do so."[92] In short, if an evolutionary process did produce the human mind, it was not Darwinian evolution. It was evolution by intelligent design.

Just as Lewis in *Miracles* rejected a Darwinian explanation for the human mind because it undermined the validity of reason, he rejected a Darwinian account of morality because it would undermine the authority of morality by attributing it to an essentially amoral process of survival of the fittest. As a practical matter, Lewis questioned whether Darwinism could actually explain the development of key human moral traits such as friendship or romantic love.[93] But in *Miracles* he made a more fundamental point: A Darwinian process "may (or may not) explain why men do in fact make moral judgments. It does not explain how they could be right in making them. It excludes, indeed, the very possibility of their being right."[94] According to Lewis, by attributing our moral beliefs and practices completely to mindless and non-moral causes, Darwinists undermined the belief that moral standards are something objectively true or even the belief that some moral beliefs are objectively preferable over others.

After all, if human behaviors and beliefs are ultimately the products of natural selection, then all such behaviors and beliefs must be equally preferable. The same Darwinian process that produces the maternal instinct also produces infanticide. The same Darwinian process that generates love also brings forth sadism. The same Darwinian process that inspires courage also spawns cowardice. Hence, the logical result of a Darwinian account of morality is not so much immorality as relativism. According to Lewis, the person who offers such an account of morality should honestly admit that "there is no such thing as wrong and right… no moral judgment can be 'true' or 'correct' and, consequently… no one system of morality can be better or worse than another."[95]

Near the end of his life, Lewis made this point with hilarious results in a "hymn" he wrote lampooning Darwinian evolution. The hymn mocked the blind and undirected nature of Darwinism: "Lead us, Evolution lead us/ Up the future's endless stair… Groping, guessing, yet progressing,/ Lead us nobody knows where." As Lewis wryly points out, once one excludes a higher purpose from biological evolution (as Darwin tried to do), traditional standards of human progress and decay no longer make sense: "Never knowing where we're going,/ We can never go astray." Applied to morality, Darwinism's philosophy of endless change repudiates "[s]tatic norms of good and evil/ (As in Plato) throned on high;/ Such scholastic, inelastic,/ Abstract yardsticks we deny."[96]

Whether in reference to man's intellect or to his morals, the cardinal difficulty with Darwinian natural selection according to Lewis is that it is mindless, and a mindless process should not be expected to produce either minds or genuine morals.

This shows why it would be so misleading to classify Lewis as a theistic evolutionist, at least as that term is typically used today. Theistic evolution can mean many things, including a form of guided evolution, but many contemporary proponents of theistic evolution are more accurately described as theistic Darwinists. That is, they do not merely advocate a guided form of common descent, but they attempt to combine evolution as an undirected Darwinian process with Christian theism. Although they believe in God, they strenuously want to avoid stating that God actually guided biological development. For example, Anglican John Polkinghorne writes that "an evolutionary universe is theologically understood as a creation allowed to make itself."[97] Former Vatican astronomer George Coyne claims that because evolution is unguided "not even God could know… with certainty" that "human life would come to be."[98] And Christian biologist Kenneth Miller of Brown University, author of the popular book *Finding Darwin's God* (which is used in many Christian colleges), insists that evolution is an undirected process, flatly denying that God guided the evolutionary process to achieve any particular result—including the development of us. Indeed, Miller insists that "mankind's appearance on this planet was not preordained, that we are here… as an afterthought, a minor detail, a happenstance in a history that might just as well have left us out."[99]

In short, many modern theistic evolutionists want to retain a belief in a Creator without actually affirming the guidance of that Creator in the history of life. In their view, the Creator delegated the development of life to a self-contained mindless process from which mind and morals emerged over time. Modern theistic evolution's attempt to strike a third way between materialism and intelligent design with a kind of emergent evolution has all the logical coherence of a circular square, or theistic atheism.

Lewis was familiar with attempts in his own day to imbue blind evolution with some sort of purposiveness while still denying the operation of a guiding intelligence, and he was not persuaded. This was where he ultimately broke with his mentor Henri Bergson. Bergson, in addition to critiquing natural selection, offered his own alternative to Darwinism, a muddled proposal for a vital force that somehow impels the evolutionary process toward integrated complexity without the need for an overarching designer. Lewis never attacked Bergson's critique of Darwinian natural selection, but after he became a Christian he repeatedly attacked Bergson's non-intelligent alternative. He did the same with George Bernard Shaw, who extolled a view, similar to to Bergson's, of "emergent evolution," the view that although evolution is not actually guided by an overarching intelligent purpose, purposeful structures that transcend blind matter somehow emerge from the process.[100]

In a section of *Mere Christianity* that is too little read, Lewis dissects this supposed third way between outright materialism and a history of life guided by design:

> People who hold this view say that the small variations by which life on this planet 'evolved' from the lowest forms to Man were not due to chance but to the 'striving' or 'purposiveness' of a Life-Force. When people say this we must ask them whether by Life-Force they mean something with a mind or not. If they do, then 'a mind bringing life into existence and leading it to perfection' is really a God, and their view is thus identical with the Religious. If they do not, then what is the sense in saying that something without a mind 'strives' or has 'purposes'? This seems to me fatal to their view.[101]

In his novel *Perelandra*, Lewis satirizes the incoherence of the emergent evolution view, which he assigns to the villain of the story, Professor E. R. Weston, a scientist run mad. Lewis gives Weston a speech of non-sequiturs and mumbo-jumbo where he solemnly appeals to "the unconsciously purposive dynamism" and "[t]he majestic spectacle of this blind, inarticulate purposiveness thrusting its way... ever upward in an endless unity of differentiated achievements toward an ever-increasing complexity of organization, towards spontaneity and spirituality." Weston ultimately identifies this blind and unconscious purposiveness with what he calls "the religious view of life" and even with "the Holy Spirit."[102]

The hero of the story, Dr. Elwin Ransom, is not impressed. "I don't know much about what people call the religious view of life," he replies. "You see, I'm a Christian. And what we mean by the Holy Ghost is not a blind, inarticulate purposiveness."[103]

Near the end of his life, Lewis read prominent theistic evolutionist Pierre Teilhard de Chardin's posthumously published book *The Phenomenon of Man*, which proposed yet another kind of emergent evolution. Lewis filled his copy of the book with critical annotations such as "Yes, he is quite ignorant," "a radically bad book," and "Ever heard of death or pain?" (The last comment responded to de Chardin's statement that "[s]omething threatens us, something is more than ever lacking, but without our being able to say exactly what."[104]) In his letters to others, Lewis called de Chardin's book "both commonplace and horrifying,"[105] and he derided de Chardin's position as "pantheistic-biolatrous waffle"[106] and "evolution run mad."[107] To a Jesuit priest Lewis even praised the Jesuits' attempt to muzzle de Chardin: "How right your Society was to shut up de Chardin!"[108]

Lewis's rejection of emergent evolution exposes why his way of thinking is ultimately so friendly to intelligent design. Lewis knew that ultimately there is no third way, no half-way house, no magical hybrid: Biological development is either the result of an unintelligent material process or a process guided by a mind, a.k.a. intelligent design. One can't split the difference. One has to choose. That being the case, Lewis thought that a mind-driven process is a far more plausible option than a mindless one.

Lewis's Critique of Evolutionism

In addition to limiting his acceptance of common descent and critiquing the power of unguided natural selection, Lewis throughout his life attacked what he called "evolutionism" or the "Myth" of "Evolution." This was evolution as a materialistic creation story that provides a competing narrative to traditional monotheism. Purporting to embody the discoveries of modern science, this "Myth" teaches that the cosmos was preceded by "the infinite void and matter endlessly, aimlessly moving to bring forth it knows not what. Then by some millionth, millionth chance—what tragic irony!—the conditions at one point of space and time bubble up into that tiny fermentation which we call organic life." Against the hostility of Nature and without purposeful direction or design, life "spreads, it breeds, it complicates itself... from the amoeba up to the reptile, up to the mammal." Finally, "there comes forth a little, naked, shivering, cowering biped, shuffling, not yet fully erect, promising nothing: the product of another millionth, millionth chance. His name in this Myth is Man." Eventually "he has become true Man. He learns to master Nature. Science arises and dissipates the superstitions of his infancy. More and more he becomes the controller of his own fate."[109] Finally, mankind becomes "a race of demi-gods" with the assistance of Darwinian eugenics, psychoanalysis, and economics. Then "the old enemy" Nature returns with a vengeance. The Sun cools, and life is "banished without hope of return from every cubic inch of infinite space. All ends in nothingness."[110]

"I grew up believing in this Myth and I have felt—I still feel—its almost perfect grandeur," observed Lewis rather wistfully. "Let no one say we are an unimaginative age: neither the Greeks nor the Norsemen ever invented a better story."[111] For Lewis, the problem with this "Myth" is not that it does not appeal to the imagination, but that it is all imagination and no logic. In fact, it contradicts the very foundation of the scientific worldview it claims to espouse.

The scientific method is premised on the idea that "rational inferences are valid," but the Myth undercuts human reason by depicting it as "simply the unforeseen and unintended by-product of a mindless process at one stage of its endless and aimless becoming. The content of the Myth thus knocks from under me the only ground on which I could possibly believe the Myth to be true." Darwin's own gnawing doubt rears its head yet again: "If my own mind is a product of the irrational... how shall I trust my mind when it tells me about Evolution?"[112]

Lewis distinguished cosmic evolutionism from the "science" of evolution, and he initially attributed it to the distortions of popularizers and journalists rather than scientists themselves. However, Lewis's distinction between evolution and evolutionism was somewhat artificial. The core of the modern scientific theory of biological evolution, after all, is Darwinism, and the core of Darwinism is the claim that evolution is an undirected material process that proceeds without either plan or foresight. Darwin himself defined natural selection as a substitute for intelligent design. In the end, then, cosmic evolutionism does not seem to be much of an extrapolation from the mainstream "scientific" theory of evolution. Indeed, the main features of what Lewis called evolutionism were baked into that scientific theory from the start.

Lewis eventually came to better understand just how intertwined evolution as a scientific theory was with what he had called evolutionism. Much of Lewis's growing awareness was likely due to his 16-year correspondence with Bernard Acworth, a leader in Britain's Evolution Protest Movement. Starting in the mid-1940s, Acworth began sending Lewis books and essays critical of Darwin's theory, materials which Lewis read and retained for his private library.[113]

Soon after coming into contact with Acworth, Lewis drew attention to a comment made by evolutionary zoologist David Watson that seemed to expose the dogmatism driving the beliefs of prominent evolutionary scientists. "Evolution," declared Prof. Watson, "...is accepted by zoologists not because it has been observed to occur or... can be proved by logically coherent evidence to be true, but because the only alternative, special creation, is clearly incredible."[114] Lewis drew this quote from an article written by two of Acworth's colleagues in the Evolution Protest

Movement. Lewis found Watson's comment "disquieting."[115] Nevertheless, he still trusted that "[m]ost biologists have a more robust belief in Evolution than Professor Watson." Otherwise it "would mean that the sole ground for believing [evolution]... is not empirical but metaphysical—the dogma of an amateur metaphysician who finds 'special creation' incredible. But I do not think it has really come to that."[116]

By 1951, Lewis was not so sure. Acworth sent him a lengthy manuscript critical of evolution, and Lewis wrote back that he had "read nearly the whole" of it. Acworth's manuscript hit home. "I must confess it has shaken me," Lewis wrote, "not in my belief in evolution, which was of the vaguest and most intermittent kind, but in my belief that the question was wholly unimportant." Lewis added that the most telling point for him was the dogmatism of the evolutionary scientists cited by Acworth: "What inclines me now to think that you may be right in regarding it [evolution] as the central and radical lie in the whole web of falsehood that now governs our lives is not so much your arguments against it as the fanatical and twisted attitudes of its defenders."[117] Lewis could no longer easily maintain that evolutionism was simply something foisted on evolutionary science by outsiders. He was appalled by the growing dogmatism and intolerance he saw among evolutionists, who seemed to treat any criticism of their views as an attack upon science itself.

Lewis had a sharply different vision of what science should be like, and he made clear that knee-jerk orthodoxy was not part of it. In Lewis's view, there was nothing anti-science about questioning dogmatic claims made in the name of science. As he came to appreciate even more deeply in the final years of his life, the scientific enterprise requires humility and an open mind in order to prosper. Those two qualities often seem sadly lacking in discussions of evolution today.

Lewis's Most Important Legacy for the Evolution Debate

"It is absolutely safe to say that if you meet somebody who claims not to believe in evolution, that person is ignorant, stupid or insane (or wicked, but I'd rather not consider that)."[118] Thus proclaims prominent evolutionary biologist Richard Dawkins from Lewis's own Oxford University. Dawkins is sometimes treated as a fringe figure because of his evangelistic atheism, but his view about the irrationality of questioning Darwinian evolution is standard fare in the evolutionary science community, where triumphalist assertions abound that the evidence for evolution is too overwhelming to question.

During Lewis's own lifetime, one finds leading evolutionary geneticist H. J. Muller declaring: "So enormous, ramifying, and consistent has the evidence for evolution become that if anyone could now disprove it, I should have my conception of the orderliness of the universe so shaken as to lead me to doubt even my own existence."[119] Or consider statements from more recent decades by evolutionary biologist Douglas Futuyma ("the statement that organisms have descended with modifications from common ancestors… is not a theory. It is a fact, as fully as the fact of the earth's revolution about the sun"[120]) and Harvard biologist Richard Lewontin ("Birds arose from nonbirds and humans from nonhumans. No person who pretends to any understanding of the natural world can deny these facts any more than she or he can deny that the earth is round, rotates on its axis, and revolves around the sun"[121]). Eugenie Scott, head of the pro-Darwin lobbying group the National Center for Science Education and someone who calls herself an "evolution evangelist," is equally emphatic: "There are no weaknesses in the theory of evolution."[122] None. Zero. Welcome to the Church of Darwinian Fundamentalism and its doctrine of scientific infallibility.

Sadly, this kind of over-the-top rhetoric is found among theistic and atheistic defenders of Darwinism alike. For example, Christian geneticist Francis Collins condemns his fellow Christians who disagree with Darwinian evolution as peddling "lies" and promoting "anti-scientific thinking."[123] Theologian Michael Peterson, meanwhile, claims that "[i]t is actually quite fair to say that evolution shares equal status with such established concepts as the roundness of the earth, its revolution around the sun, and the molecular composition of matter."[124] Get the point?

The person who criticizes Darwin's theory is equivalent to someone who thinks the earth is flat, who believes the sun revolves around the earth, and who apparently doesn't accept microscopes or the periodic table.

It is hard to believe that Lewis would have had any sympathy at all for this kind of bluster. After all, he himself questioned large chunks of modern evolutionary theory, including the ability of natural selection to explain mind, morality, and the development of complex biological structures. Lewis did grant that biological evolution was a "genuine scientific hypothesis" worthy of discussion.[125] But he sharply distinguished in his own mind a "hypothesis" from dogmatic claims that something is a "basic fact." Lewis was very clear that what he meant by "hypothesis" was an interpretation of facts based on assumptions; and a hypothesis must therefore always be open to challenge and repeal. In his view "real biologists" (as opposed to propagandists) recognized that evolution is simply a hypothesis, not a dogmatic truth. "It covers more of the facts than any other hypothesis at present on the market and is therefore to be accepted unless, or until, some new supposal can be shown to cover still more facts with even fewer assumptions."[126]

At the root of Lewis's willingness to question evolutionary claims was a healthy skepticism of the scientific enterprise itself. Lewis respected modern science, and he respected modern scientists. But unlike many contemporary champions of evolution, he did not embrace a simple-minded view of natural science as fundamentally more authoritative or less prone to error than all other fields of human endeavor.

One of the last books about science Lewis read before he died was *The Open Society and Its Enemies* by philosopher Karl Popper. Near the end of that book, Popper frankly admits the lack of objectivity to be found even in experimental science. Lewis underlined the passage:

> For even our experimental and observational experience does not consist of 'data'. Rather, it consists of a web of guesses—of conjectures, expectations, hypotheses with which there are interwoven accepted, traditional, scientific, and unscientific, lore and prejudice. There simply is

no such thing as pure experimental and observational experience—experience untainted by expectation and theory.[127]

Lewis's growing awareness of the human fallibility of science was expressed powerfully in his final book, *The Discarded Image* (1964).[128] Published after his death, the book is ostensibly about the medieval worldview. But the nature of science is one of the underlying themes. Lewis argues in the book that scientific theories are "supposals" and should not be confused with "facts." Properly speaking, scientific theories try to account for as many facts as possible with as few assumptions as possible. But according to Lewis, we must always recognize that such explanations can be wrong: "In every age it will be apparent to accurate thinkers that scientific theories, being arrived at in the way I have described, are never statements of fact. That stars appear to move in such and such ways, or that substances behaved thus and thus in the laboratory—these are statements of fact."[129] By contrast, the theories that seek to explain these facts "can never be more than provisional." They "have to be abandoned" if someone thinks of a "supposal" that can account for "observed phenomena with still fewer assumptions, or if we discover new phenomena" that the previous theory cannot account for "at all."[130] Lewis said he believed that "all *thoughtful* scientists today" would be able to recognize this truth, although he did not speculate about how many "thoughtful scientists" actually exist. He did think the biggest problem with scientific dogmatism lay outside the scientific community, where "[t]he mass media… have in our time created a popular scientism, a caricature of the true sciences."[131] Nevertheless, any scientist who engages in such dogmatism would clearly be doing something inappropriate according to Lewis.

However, the truly radical part of Lewis's critique of modern science was still to come. In his epilogue to *The Discarded Image*, Lewis discusses at length the shift from the medieval to the modern model of biology. It soon becomes evident that he does not think empirical evidence drives scientific revolutions. Lewis declares that the Darwinian revolution in particular "was certainly not brought about by the discovery of new facts."[132]

Lewis recalled that when he was young he "believed that 'Darwin discovered evolution' and that the far more general, radical, and even cosmic developmentalism... was a superstructure raised on the biological theorem. This view has been sufficiently disproved." What really happened according to Lewis was that "[t]he demand for a developing world—a demand obviously in harmony both with the revolutionary and the romantic temper" had developed first, and when it was "full grown" the scientists went "to work and discover[ed] the evidence on which our belief in that sort of universe would now be held to rest."[133]

Lewis's view has momentous implications for how we view the reigning paradigms in science at any given time—including Darwinian evolution. "We can no longer dismiss the change of Models [in science] as a simple progress from error to truth," argued Lewis. "No Model is a catalogue of ultimate realities, and none is a mere fantasy... But... each reflects the prevalent psychology of an age almost as much as it reflects the state of that age's knowledge." Lewis added that he did "not at all mean that these new phenomena are illusory... But nature gives most of her evidence in answer to the questions we ask her."[134]

So the answers we receive from nature are dictated by the questions we ask, and the questions we ask are shaped by the assumptions and expectations of the scientific theories we embrace—assumptions and expectations likely borrowed from larger cultural attitudes that predated the scientific evidence they seek to interpret. Hence, the potential for even good scientific theories to blind us to key aspects of reality is huge.

Nowhere is this more true than in the field of Darwinian evolution itself, which is based on the inviolable assumption that everything in biology must be the result of unguided material processes. Over the past century, this assumption has undoubtedly inspired many interesting research questions and scientific advances. At the same time, it also has undoubtedly discouraged and delayed many *other* important research questions. Witness the unhelpful Darwinian preoccupation with "vestigial" organs over the past century. Time and again, biological features we do not fully understand have been dismissed by advocates of Darwinian evolution as non-functional leftovers from a blind evolutionary process. Time and again, researchers who eventually bothered to look discovered that such supposedly "vestigial" features—the appendix and tonsils, to name two—actually perform important biological functions.[135] The evidence of function was there all along, but scientists were discouraged by the existing paradigm from asking the questions that would elicit the evidence.

More recently, one of the biggest mistakes in the history of modern biology may turn out to be the belief that the human genome is riddled with "junk DNA." Random mutations in protein-coding DNA are supposed to drive Darwinian evolution, and so when it was discovered that the vast majority of DNA does not code for proteins, some leading Darwinists jumped to the conclusion that non-protein-coding DNA must be mere "junk" left over from the evolutionary process just like some vestigial organs. Not only that, leading evolutionists ranging from atheist Richard Dawkins to Christian Francis Collins championed "junk DNA" as proof that human beings were the result of a Darwinian process rather than intentional design.[136]

However, when scientists finally started to look more closely at non-coding DNA, they were shocked to learn that reality did not correspond to their ideological assumptions. Indeed, over the past decade science journals have been flooded with new research showing the rich and varied functionality of so-called "junk DNA." In the words of biologist Jonathan Wells: "Far from consisting mainly of junk that provides evidence against intelligent design, our genome is increasingly revealing itself to be a multidimensional, integrated system in which non-protein-coding DNA performs a wide variety of functions."[137] Again, the evidence of functionality in non-protein-coding DNA was always there to find; but the evidence was not forthcoming because few people were asking the right questions. As Lewis pointed out so perceptively, treating reigning paradigms in science as all-encompassing dogmas will blind us to how much about nature we may be missing. Such dogmatism also breeds a kind of scientific authoritarianism that is incompatible with a free society, which Lewis eloquently rebuked in books such as *The Abolition of Man* and *That Hideous Strength*.[138]

By highlighting the all-too-human frailties of modern science, Lewis made his most important contribution to the evolution debate. In essence, Lewis legitimized the right to dissent from Darwin. By stressing the non-scientific underpinnings of scientific revolutions, Lewis showed that Darwinian evolution should not be privileged as some special form of knowledge that is immune from critical scrutiny. By exposing just how limited a window on reality a given scientific theory can provide, he validated the continued questioning of Darwinian evolution as well as other theories in science.

Indeed, Lewis predicted that it was partly by raising the right questions that the current (Darwinian) model of biology might be replaced. He used the analogy of placing someone on trial: "Here, as in the courts, the character of the evidence depends on the shape of the examination, and a good cross-examiner can do wonders."[139]

Lewis's words proved prophetic. In 1991, Berkeley law professor Phillip Johnson did precisely what Lewis had described by publishing his book *Darwin on Trial*, which mounted a full-throttled cross-examination of the standard evidence for orthodox Darwinism.[140] C. S. Lewis was proved right again: A "good cross-examiner" really *can* "do wonders." Igniting a furor among leading Darwinists, Johnson's book helped inspire a whole new generation of scientists and philosophers to launch increasingly sophisticated challenges to Darwinian theory as well as to formulate a fresh argument for intelligent design in nature.

Like Lewis, Phillip Johnson understood that "nature gives most of her evidence in answer to the questions we ask her." And he recognized the critical importance of asking "the right questions" in scientific debates—even when those questions may make the guardians of the existing paradigm uncomfortable or angry.[141]

Those who truly want to honor C. S. Lewis's legacy in the area of science and society would do well to do likewise.[142]

Endnotes

1. For good accounts of the real story of the Scopes trial and the inaccuracies of *Inherit the Wind*, see Larson, *Summer for the Gods* (New York: Basic Books, 1997); Carol Iannone, "The Truth about Inherit the Wind," *First Things*, February 1997, accessed May 18, 2012, http://www.firstthings.com/article.php3?id_article=3645; Marvin Olasky and John Perry, *Monkey Business: The True Story of the Scopes Trial* (Nashville: Broadman and Holman, 2005).
2. See George Sayer, *Jack: A Life of C. S. Lewis* (Wheaton, IL: Crossway, 1994), 176–184; Roger Lancelyn Green and Walter Hooper, *C. S. Lewis: A Biography*, rev. edition (San Diego: Harcourt Brace & Company, 1994), 79–85.
3. C. S. Lewis to his Father, Aug. 14, 1925 in *C. S. Lewis: Collected Letters*, edited by Walter Hooper (London: HarperCollins, 2000), vol. I, 649.
4. Francis Collins, *The Language of God* (New York: Free Press, 2006), 21–31, 208–209, 222–225.
5. Michael L. Peterson, "C. S. Lewis on Evolution and Intelligent Design," *Perspectives on Science and Christian Faith* 62, no. 4 (December 2010): 253–266.
6. Michael L. Peterson, "C. S. Lewis on Evolution and Intelligent Design," www.biologos.org (April 2011), accessed May 18, 2012, http://biologos.org/uploads/projects/peterson_scholarly_essay.pdf.
7. David Roach, "Poll: Pastors Oppose Evolution, Split on Earth's Age," www.lifeway.com, Jan. 9, 2012, accessed May 18, 2012, http://www.lifeway.com/Article/Research-Poll-Pastors-oppose-evolution-split-on-earths-age.
8. Kenton Sparks, "After Inerrancy: Evangelicals and the Bible in a Postmodern Age," BioLogos Forum, June 2010, accessed May 18, 2012, http://biologos.org/uploads/static-content/sparks_scholarly_essay.pdf.
9. Karl Giberson, *Saving Darwin* (New York: HarperOne, 2008), 11–13.
10. Miller quoted in John G. West, *Darwin Day in America: How Our Politics and Culture Have Been Dehumanized in the Name of Science* (Wilmington, DE: ISI Books, 2007), 226; more generally, see discussion on 225–227; also see Kenneth Miller, *Finding Darwin's God* (New York: HarperCollins, 1999).
11. Some of these books and pamphlets with a significant focus on evolution in Lewis's library include Bernard Acworth, *The Cuckoo*; Alfred Balfour, *Theism and Humanism*; Henri Bergson, *L'Evolution Creatrice*; Charles Darwin, *Autobiography*; L. M. Davies, *BBC Abuses Its Monopoly*; L. M. Davies, *Evolutionists Under Fire*; Douglas Dewar, *The Man from Monkey Myth*; Douglas Dewar, *Science and the BBC*; Evolution Protest Movement, *Evolution: How the Doctrine Is Propagated in Our Schools*; C. W. Formby, *The Unveiling of the Fall*; E. O. James, *Evolution and the Fall*; Edmund Sinnott, *Biology of the Spirit*; Joseph Solomon, *Bergson*; Pierre Teilhard de Chardin, *The Phenomenon of Man*. Much of Lewis's personal library is currently housed at the Wade Center, Wheaton College. For a listing of the extant books from Lewis's personal library, consult the description in "C. S. Lewis Library" (Wade Center, 2010), accessed May 18, 2012, http://www.wheaton.edu/wadecenter/Collections-and-Services/Collection%20Listings/~/media/Files/Centers-and-Institutes/Wade-Center/RR-Docs/Non-archive%20Listings/Lewis_Public_shelf.pdf.

12. C. S. Lewis to Dom Bede Griffiths, July 5, 1949, *The Collected Letters of C. S. Lewis* (San Francisco: HarperSanFrancisco, 2004), vol., II, 953.
13. C. S. Lewis to Bernard Acworth, Oct. 4, 1951, in *The Collected Letters of C. S. Lewis*, edited by Walter Hooper (San Francisco: HarperSanFrancisco, 2007), vol. III, 140.
14. Peterson, "C. S. Lewis on Evolution and Intelligent Design" (2010), 264.
15. C. S. Lewis, *Mere Christianity* (New York: Macmillan, 1960), 154.
16. A written transcript of the radio talk that formed the basis of this part of *Mere Christianity* can be found at "C. S. Lewis—Only Surviving Episode of Broadcast Talks," accessed May 19, 2012, http://www.awesomestories.com/assets/cs-lewis-only-surviving-episode-of-broadcast-talks. The website also has the original audio for this broadcast talk by Lewis.
17. On the different meanings of evolution, see the helpful discussion by Jay Richards, editor, *God and Evolution: Protestants, Catholics, and Jews Explore Darwin's Challenge to Faith* (Seattle: Discovery Institute Press, 2010), 18–25.
18. C. S. Lewis to Bernard Acworth, Dec. 9, 1944, *Collected Letters*, vol. II, 633.
19. C. S. Lewis to Joseph Canfield, Feb. 28, 1955, unpublished letter, Wade Center Collection, Wheaton College.
20. C. S. Lewis, *The Problem of Pain* (New York: Macmillan, 1962), 76–77.
21. C. S. Lewis to Sister Penelope, Jan. 10, 1952, *Collected Letters*, vol. III, 157.
22. C. S. Lewis to Bernard Acworth, Sept. 23, 1944, reprinted in Gary B. Ferngren and Ronald L. Numbers, "C. S. Lewis on Creation and Evolution: The Acworth Letters, 1944–1960," *Perspectives on Science and Christian Faith* 48 (March 1996), accessed May 18, 2012, http://www.asa3.org/ASA/PSCF/1996/PSCF3-96Ferngren.html.
23. For more information about the Piltdown hoax, see Frank Spencer, *Piltdown: A Scientific Forgery* (New York: Oxford University Press, 1990).
24. C. S. Lewis to Bernard Acworth, Dec. 16, 1953, reprinted in Ferngren and Numbers, "C. S. Lewis on Creation and Evolution."
25. C. S. Lewis to Joseph Canfield, Feb. 28, 1955.
26. C. S. Lewis, "The Adam Unparadised" in *Poems*, edited by Walter Hooper (San Diego: 1964), 44. This poem was originally published in September 1949.
27. See C. S. Lewis, *The Last Battle* (New York: Macmillan, 1956).
28. Interview with Bishop John Shelby Spong, *Compass* [television program on ABC network in Australia], July 8, 2001, accessed May 18, 2012, http://www.abc.net.au/compass/intervs/spong2001.htm.
29. Giberson, *Saving Darwin*, 12.
30. Ibid.
31. C. S. Formby, *The Unveiling of the Fall* (London: Williams and Norgate, 1923).
32. Ibid., 28.
33. Ibid., 30.
34. Ibid., 33.
35. Ibid., xiv–xxiii, 34, 93–109, 177–187.
36. Lewis, *Problem of Pain*, 133–136.

37. C. S. Lewis to Bernard Acworth, Sept. 23, 1944.
38. Lewis, *Problem of Pain*, 88.
39. Ibid., 78–79.
40. Ibid., 79.
41. Ibid., 78.
42. J. R. R. Tolkien to Christopher Tolkien, Jan. 30, 1945, in *The Letters of J. R. R. Tolkien*, selected and edited by Humphrey Carpenter (Boston: Houghton Mifflin Company, 1981), 109–110.
43. Ibid., 110, emphasis in original.
44. Ibid., 109.
45. Lewis, *Problem of Pain*, 74.
46. Ibid.
47. See discussion of Darwin's account of morality in John G. West, *Darwin Day in America*, 29–37.
48. Brian D. McLaren, *A New Kind of Christianity: Ten Questions that Are Transforming the Faith*, Kindle edition (HarperCollins E-Books, 2010), 49–50.
49. C. S. Lewis, *Perelandra* (New York: Macmillan, 1965).
50. C. S. Lewis, *Miracles: A Preliminary Study* (New York: Macmillan, 1947), 95.
51. C. S. Lewis to Joseph Canfield, Feb. 28, 1955.
52. C. S. Lewis to Miss Jacob, July 3, 1951, unpublished letter, Wade Center Collection, Wheaton College.
53. Tim Keller, "Creation, Evolution and Christian Laypeople," www.biologos.org (February 2011): 7, accessed May 19, 2012, http://biologos.org/uploads/projects/Keller_white_paper.pdf.
54. In his diary for Aug. 18, 1925, Lewis relates that Maureen Moore asked him how Adam and Eve related to evolutionary theory, and he replied that "the Biblical and scientific accounts were alternatives. She asked me which I believed. I said the scientific." *All My Road Before Me: The Diary of C. S. Lewis, 1922–1927*, edited by Walter Hooper (San Diego: Harcourt Brace Jovanovich, 1991), 361. This was the same month Lewis expressed his doubts about the ideas of Darwin and Spencer to his father.
55. Lewis, *Problem of Pain*, 79.
56. A. N. Wilson, *C. S. Lewis: A Biography* (New York: W.W. Norton, 1990), 210.
57. Lewis, *Perelandra*, 120.
58. C. S. Lewis to Don Giovanni Calabria, March 17, 1953, *Collected Letters*, vol. III, 306.
59. C. S. Lewis to Mrs. Hook, Dec. 29, 1958, *Collected Letters*, vol. III, 1004.
60. C. S. Lewis to Sister Penelope, Jan. 10, 1952.
61. Charles Darwin, *The Autobiography of Charles Darwin and Selected Letters*, edited by F. Darwin (New York: Dover Publications, 1958), 63.
62. Charles Darwin, *The Variation of Animals and Plants under Domestication*, second edition (London: John Murray, 1875), vol. II, 428–429.

63. Letter from Nobel Laureates to Kansas State Board of Education, September 9, 2005, accessed May 18, 2012, http://media.ljworld.com/pdf/2005/09/15/nobel_letter.pdf.
64. C. S. Lewis, "Modern Man and His Categories of Thought," *Present Concerns*, edited by Walter Hooper (New York: Harcourt Brace Jovanovich, 1986), 63.
65. C. S. Lewis, "The Funeral of a Great Myth," in *Christian Reflections*, edited by Walter Hooper (Grand Rapids, MI: Eerdmans, 1967), 86.
66. C. S. Lewis, "Is Theology Poetry?" in *The Weight of Glory and Other Addresses*, rev. and expanded edition edited by Walter Hooper (New York: Macmillan, 1980), 89.
67. Henri Bergson, *Creative Evolution*, translated by Arthur Mitchell (London: Macmillan, 1920).
68. C. S. Lewis, *Surprised by Joy* (New York: Harcourt Brace Jovanovich 1955), 198.
69. C. S. Lewis to Arthur Greeves, June 19, 1920, *Collected Letters*, vol. I, 494.
70. C. S. Lewis to his father, Sept. 13, 1919, *Collected Letters*, vol. I, 464.
71. Lewis's annotated copy of *L'Evolution Creatice* resides in the Wade Center Collection, Wheaton College.
72. Bergson, *Creative Evolution*, 59.
73. Ibid., 68.
74. Ibid., 69–70.
75. Ibid., 57.
76. Ibid., 59–60.
77. C. S. Lewis, annotation to copy of Henri Bergson, *L'Evolution Creatice* (Paris, 1917), 60; Wade Center Collection, Wheaton College.
78. Ibid., 61.
79. Ibid., 74.
80. C. S. Lewis to his father, Aug. 14, 1925.
81. G. K. Chesterton, *The Everlasting Man* (San Francisco: Ignatius, 1993), 42.
82. Ibid., 38.
83. Arthur J. Balfour, *Theism and Humanism*, edited by Michael W. Perry (Seattle: Inkling Books, 2000), 138.
84. Ibid., 141.
85. C. S. Lewis, "Is Theology Poetry?" 77–78. Albert Lewis's copy of *Theism and Humanism* currently resides in the Wade Center Collection, Wheaton College.
86. In answer to a query from *The Christian Century* magazine, Lewis listed *Theism and Humanism* as one of ten books that "did most to shape" his "vocational attitude" and his "philosophy of life." Lewis's list was published in the June 6, 1962 edition of the magazine.
87. Paul F. Ford, "Arthur James Balfour," in Jeffrey D. Schultz and John G. West, editors, *The C. S. Lewis Readers' Encyclopedia* (Grand Rapids, MI: Zondervan, 1998), 92.
88. C. S. Lewis, *Miracles: A Preliminary Study*, 1960 edition (New York: Macmillan, 1978), 18.
89. Ibid., 18–19.

90. C. S. Lewis's copy of Charles Darwin, *Autobiography of Charles Darwin*, The Thinker's Library No. 7 (London: Watts & Co. for the Rationalist Press, 1929), 149. Wade Center Collection, Wheaton College.
91. Ibid., 153.
92. Lewis, *Miracles* (1960 edition), 22–23.
93. C. S. Lewis, *The Four Loves* (New York: Harcourt Brace Jovanovich, 1960), 90.
94. Lewis, *Miracles* (1960 edition), 36.
95. Ibid.
96. "Evolutionary Hymn," in Lewis, *Poems*, 55.
97. John Polkinghorne, *Quarks, Chaos, and Christianity* (New York: Crossroad Publishing Company, 2005), 113.
98. George V. Coyne, S.J., "The Dance of the Fertile Universe" (2005): 7, formerly available at http://www.aei.org/docLib/20051027_HandoutCoyne.pdf.
99. Miller, *Finding Darwin's God*, 272.
100. See Lewis, *Mere Christianity*, 35; C. S. Lewis to Bernard Acworth, March 5, 1960, *Collected Letters*, vol. III, 137; Lewis, *The Four Loves*, 152–153; C. S. Lewis, *Studies in Words* (Cambridge: Cambridge University Press, 1960), 300–301.
101. Lewis, *Mere Christianity*, 35.
102. Lewis, *Perelandra*, 90–91.
103. Ibid., 91.
104. C. S. Lewis, annotations to Pierre Teilhard de Chardin, *The Phenomenon of Man*, with an introduction by Sir Julian Huxley (London: Collins, 1959), 217, cover page, 227. Wade Center Collection, Wheaton College.
105. C. S. Lewis to Dan Tucker, Dec. 8, 1959, *Collected Letters*, vol. III, 1105.
106. C. S. Lewis to Father Frederick Joseph Adelmann S.J., Sept. 21, 1960, *Collected Letters*, vol. III, 1186.
107. C. S. Lewis to Bernard Acworth, March 5, 1960, *Collected Letters*, vol. III, 1137.
108. C. S. Lewis to Father Frederick Joseph Adelmann S.J., Sept. 21, 1960.
109. Lewis, "The Funeral of a Great Myth," 87.
110. Ibid., 88.
111. Ibid.
112. Ibid., 89.
113. These materials included Bernard Acworth, *The Cuckoo*; L. M. Davies, *BBC Abuses Its Monopoly*; L. M. Davies, *Evolutionists Under Fire*; Douglas Dewar, *The Man from Monkey Myth*; Douglas Dewar, *Science and the BBC*; and Evolution Protest Movement, *Evolution: How the Doctrine Is Propagated in Our Schools*. All of these materials reside in the Wade Center Collection, Wheaton College.
114. C. S. Lewis, "Is Theology Poetry?" 89 and "Funeral of a Great Myth," 85.
115. Lewis, "Is Theology Poetry?" 89.
116. Lewis, "Funeral of a Great Myth," 85.
117. C. S. Lewis to Bernard Acworth, Sept. 13, 1951, *Collected Letters*, vol. III, 138.

118. Richard Dawkins, "Put Your Money on Evolution," *The New York Times*, April 9, 1989, section VII, 35.

119. H. J. Muller, quoted in J. Peter Zetterberg, editor, *Evolution Versus Creationism: The Public Education Controversy* (Oryx Press, 1983), 33–34.

120. Douglas J. Futuyma, *Evolutionary Biology*, second edition (Sunderland, MA: Sinauer Associates, 1986), 15.

121. Richard Lewontin, quoted in Zetterberg, *Evolution Versus Creationism*, 31.

122. Eugenie Scott, quoted in Ed Stoddard, "Evolution gets added boost in Texas schools," Reuters.com, accessed May 19, 2012, http://blogs.reuters.com/faith-world/2009/01/23/evolution-gets-added-boost-in-texas-schools/. For Eugenie Scott's description of herself as an "evolution evangelist," see Ronald L. Numbers, *The Creationists: From Scientific Creationism to Intelligent Design*, expanded edition (Cambridge, MA: Harvard University Press, 2006), 380.

123. Francis Collins, "Foreword" to Giberson, *Saving Darwin*, v, vii.

124. Michael Peterson, "C. S. Lewis on Evolution and Intelligent Design" (2010), n. 29, 266.

125. Lewis, "Funeral of a Great Myth," 83.

126. Ibid., 85.

127. C. S. Lewis's copy of K. R. Popper, *The Open Society and Its Enemies*, vol. II—*The High Tide of Prophecy: Hegel, Marx, and the Aftermath* (London: Routledge & Kegan Paul), 388. Wade Center Collection, Wheaton College.

128. C. S. Lewis, *The Discarded Image* (Cambridge: Cambridge University Press, 1964).

129. Ibid., 15–16.

130. Ibid.

131. Ibid., 16, 17

132. Ibid., 220.

133. Ibid., 220–221.

134. Ibid., 222–223.

135. See Casey Luskin, "Vestigial Arguments about Vestigial Organs Appear in Proposed Texas Teaching Materials," Evolution News and Views, June 20, 2011, accessed May 19, 2012, http://www.evolutionnews.org/2011/06/vestigial_arguments_about_vest047341.html; David Klinghoffer, "Looks Like the Appendix Isn't a 'Junk Body Part' After All," Evolution News and Views, Jan. 4, 2012, accessed May 19, 2012, http://www.evolutionnews.org/2012/01/now_its_the_app054761.html.

136. Jonathan Wells, *The Myth of Junk DNA* (Seattle: Discovery Institute Press, 2011), 19–20, 23–24, 98–100. In more recent years, Collins has seemed to abandon or at least diminish his support for the junk DNA paradigm. See ibid., 98–100.

137. Ibid., 9.

138. C. S. Lewis, *The Abolition of Man* (New York: Macmillan, 1955); *That Hideous Strength* (New York: Macmillan, 1965).

139. Lewis, *The Discarded Image*, 223.

140. Phillip E. Johnson, *Darwin on Trial*, second edition (Downers Grove, IL: InterVarsity Press, 1993).

141. See Phillip E. Johnson, *The Right Questions* (Downers Grove, IL: InterVarsity, 2002).
142. I gratefully acknowledge the research of Jake Akins at the Wade Center Collection, Wheaton College, which contributed to this essay, and the permission of both the Wade Center and the C. S. Lewis Company to quote from some of Lewis's unpublished writings deposited at the Wade Center. Finally, I would like to thank Jay Richards, Sonja West, and Cameron Wybrow for their thoughtful comments on an earlier draft of this essay.

7

C. S. LEWIS AND INTELLIGENT DESIGN

John G. West

A FEW MONTHS AFTER BEING DISCHARGED FROM THE ARMY AT THE end of World War I, a twenty-year-old C. S. Lewis published his first book, a cycle of poems titled *Spirits in Bondage* (1919).[1] The opening poem, "Satan Speaks," provided a grim view of nature that might startle many of Lewis's later readers:

> I am Nature, the Mighty Mother,
> I am the law: ye have none other.
> I am the flower and the dewdrop fresh,
> I am the lust in your itching flesh.
> I am the battle's filth and strain,
> I am the widow's empty pain.
> I am the sea to smother your breath,
> I am the bomb, the falling death…
> I am the spider making her net,
> I am the beast with jaws blood-wet[2]

Many of the poems in *Spirits in Bondage* displayed Lewis's passionate, even angry, atheism during this period of his life, inspired by what he referred to as the "Argument from Undesign": the idea that the pain, cruelty, and wastefulness of nature supplies the best evidence against the view that a benevolent deity created the world.[3]

Lewis's captivation by the argument from undesign reflected not only the personal tragedies of his life (such as the death of his mother from cancer during his childhood), but also his reading of scientific materialists like H. G. Wells who nourished the young Lewis's imagination with depictions of the universe as vast, cold, and impersonal.[4] Even after Lewis became a Christian, he remained skeptical of the traditional "argument from design," which purported to show that from various features of nature one can prove the existence of the Christian God. In his book *The Problem of Pain* (1940), Lewis prefaces his first chapter with a comment by Pascal that "[i]t is a remarkable fact that no canonical writer has ever used Nature to prove God."[5] Writing to a former student in 1946, Lewis reaffirmed that "I still think the argument from design the weakest possible ground for Theism, and what may be called the argument from un-design the strongest for Atheism."[6] And writing in his autobiography *Surprised by Joy* (1956), Lewis recalled the impact on his thought of the Greek materialist Lucretius's refutation of design: "Had God designed the world, it would not be / A world so frail and faulty as we see."[7]

Lewis's powerful attraction to the argument from undesign makes all the more remarkable his eventual embrace of several arguments favorable to design.

Four Lewis Arguments Friendly to a Universe by Design

Lewis countered the argument from undesign with several positive arguments in favor of the existence of a transcendent intelligent cause for nature. They include:

1. The argument from natural beauty.

From early on, Lewis's pessimistic view of nature as "red in tooth and claw"[8] was counterbalanced by the longings stirred within him by nature's beauty.[9] Even in *Spirits in Bondage*, the bleak vision of nature presented in some of his poems can be contrasted with poems describing scenes of overwhelming beauty that raised glimmers of the transcendent. For Lewis, our experience of beauty in nature pointed to the reality of something beyond nature:

Atoms dead could never thus
Stir the human heart of us
Unless the beauty that we see
The veil of endless beauty be[10]

In Lewis's view, the longings provoked by earthly beauty could not be accounted for by a blind and mechanistic material universe. They required a transcendent cause outside of nature. This cause was not necessarily personal, but it did go beyond blind matter and energy. As a consequence, it put an intelligent agent back on the table as one of the options for discussion.

2. The argument from morality.

LEWIS EVENTUALLY recognized that the argument from undesign suffered from a critical flaw: If the material universe is all there is, and if human beings are simply the products of that universe, then on what basis can they criticize the universe for being so bad?[11] By judging the universe in this way, human beings are presupposing the existence of a moral standard outside of material nature that can judge nature. But where did this moral standard come from? The existence in every culture of a standard by which the current operations of nature are judged implies the existence of a transcendent moral cause outside of nature. Again, this transcendent moral cause is not necessarily personal, but a transcendent personal God is one of the alternatives that can now be considered.

3. The argument from reason.

As DISCUSSED in detail in chapters 6, 8, and 9, Lewis argued that reason cannot be accounted for by an undirected material process of chance and necessity such as natural selection acting on random mutations. If reason could be accounted for in this way, according to Lewis, we would have no reason to trust the conclusions of our minds, including the conclusion that our minds are the products of a material process of chance and necessity. The bottom line for Lewis is that the existence of reason within nature points to a need for reason outside of nature as a transcendent intelligent cause.¹²

4. The argument from functional complexity.

ACCORDING TO Lewis, "universal evolutionism" has schooled us to think that in nature complicated functional things naturally arise from cruder and less complicated things. Oak trees come from acorns, owls from eggs, and human beings from embryos. But for Lewis this "modern acquiescence in universal evolutionism is a kind of optical illusion" that defies the actual data of the natural world.¹³ In each of the aforementioned cases, complex living things arose from even more complex living things. Every acorn originally came from an oak tree. Every owl's egg came from an actual owl. Every human embryo required two full-grown adult human beings. We see the same pattern in human culture. The "evolution" from coracles to steamships, or from one of the early locomotives (the "Rocket") to modern train engines, requires a cause that is greater than either steamships or train engines. "We love to notice that the express [train] engine of today is the descendant of the 'Rocket'; we do not equally remember that the 'Rocket' springs not from some even more rudimentary engine, but from something much more perfect and complicated than itself—namely, a man of genius."¹⁴ Lewis made clear the relevance of this truth for understanding the wonderful functional complexity we see throughout nature: "You have to go outside the sequence of engines, into the world of men, to find the real originator of the Rocket. Is it not equally reasonable to look outside Nature for the real Originator of the natural order?"¹⁵

This is explicitly an argument for intelligent design, and Lewis implies that this line of reasoning was central to his own disavowal of materialism. "On these grounds and others like them one is driven to think that whatever else may be true, the popular scientific cosmology at any rate is certainly not."¹⁶ This argument for intelligent design does not in and of itself lead to the Christian God according to Lewis. But it opens the door to considering the alternatives to materialism of "philosophical idealism" and "Theism," and from there one may well progress to full-blooded Christian theism after further reflection.[17]

Despite Lewis's clear support for the idea that key features of nature point to an intelligent cause, there have been efforts recently to cast Lewis as an opponent of the contemporary argument for intelligent design. Most notably, theologian Michael Peterson argued this thesis in a lengthy piece published in 2010 and serialized the following year on the BioLogos Foundation website.[18] Peterson's mischaracterization of Lewis's views on Darwinian evolution were already discussed in chapter 6. The next section will examine Peterson's misrepresentation of Lewis on the question of intelligent design, followed by a discussion of how Lewis in fact rebutted several key objections raised against the modern theory of intelligent design. (Readers more interested in what Lewis had to say than in Peterson's interpretations of Lewis are invited to skip the following section.)

Lewis and the Straw Man Version of Intelligent Design

MICHAEL PETERSON ACKNOWLEDGES THE OBVIOUS fact that Lewis put forward various arguments in favor of an intelligent cause of the universe, but then asserts that "none of these lines of reasoning are really design-type arguments."[19] This statement is clearly false with regard to the argument from functional complexity described above (which Peterson does not discuss); but Peterson's claim is also misleading with regard to Lewis's argument from reason, which Peterson himself earlier admits is "closely related" to "design-type" arguments. Regardless, Peterson's overall point is even more problematic: His thesis seems to be that Lewis would have rejected the argument for intelligent design as it has been developed over the past several years by scientists and philosophers such as Michael Behe, William Dembski, and Stephen Meyer.

Peterson's evidence for this thesis is thin to say the least. Lewis died in 1963, and so any arguments about whether Lewis "would" have opposed or embraced current arguments about intelligent design are highly speculative. Aslan's warning in the Chronicles of Narnia would seem to be apropos: "[N]o one is ever told *what would have happened.*"[20] Be that as it may, the most serious problem with Peterson's thesis is that it is based more on his misunderstanding of intelligent design than it is on Lewis's views about anything. Indeed, Peterson spends more than 40% of his article providing a highly inaccurate rendition of intelligent design rather than presenting evidence of Lewis's own views on the topic. Now if intelligent design really consisted of some of the claims Peterson puts forward, Lewis might well have opposed it. The problem is that Peterson misrepresents the modern theory of intelligent design, and so his article in the end provides almost no insight into what Lewis would have thought about the theory as espoused by its actual proponents.

Unfortunately, Peterson shows scant evidence that he has read much by the intelligent design proponents he seeks to critique. Of the scores of books and technical articles published by intelligent design theorists over the past two decades, Peterson appears to base his criticisms on a grand total of two books, Michael Behe's *Darwin's Black Box* (1996) and William Dembski's *Intelligent Design: The Bridge Between Science and Theology* (1999).[21] Both volumes were published more than a decade ago. Although important, they hardly provide an adequate summary of the current state of thinking among intelligent design proponents. In a footnote to his original article, Peterson does reference a third book, *Signature in the Cell*, published in 2009, but it is unclear whether he has read it since he attributes the book to someone named "David Myer." In fact, the book was written by Cambridge-trained philosopher of science Stephen Meyer.[22]

Peterson's mischaracterization of intelligent design commences with his initial description of its origins. He states that "[i]n the late 1990s, the 'intelligent design' (ID) movement emerged… rejecting evolutionary principles and purporting to have a hot, new scientific argument for God."[23] There are multiple problems with this claim.

First, although the modern intelligent design movement gained prominence by the late 1990s, it began to emerge considerably earlier with the publication of the books *Chance or Design?* in 1979, *The Mystery of Life's Origin* in 1984, and *Evolution: A Theory in Crisis* in 1985.[24] These books, in turn, built upon discoveries in physics, cosmology, and biology reaching back to the 1950s.[25]

Second, intelligent design theorists did not offer intelligent design as "a hot, new scientific argument for God." In fact, leading intelligent design proponents (especially William Dembski and Michael Behe, the two scholars cited by Peterson) made a much more nuanced claim. They argued that scientific evidence corroborates the proposition that key features of nature are the product of an intelligent cause rather than an undirected process such as Darwinian natural selection. Far from claiming that this is a "hot, new scientific argument for God," they emphasized that modern intelligent design theory in biology—unlike the old natural theology of say, William Paley—could not reach that far on its own. This is not to deny that intelligent design has positive implications for belief in God. If nature supplies evidence that it is the product of an intelligently guided process, that understanding of reality certainly lends more support to belief in God than the idea that nature is the product of a blind, undirected process like Darwinism. Nevertheless, any claim that an intelligent cause detected by science must be God (let alone the God of the Bible) requires additional arguments from philosophy and metaphysics to justify it.

In the words of biochemist Michael Behe, intelligent design as a scientific research program "is limited to design itself; I strongly emphasize that it is not an argument for the existence of a benevolent God, as Paley's was. I hasten to add that I myself do believe in a benevolent God, and I recognize that philosophy and theology may be able to extend the argument. But a scientific argument for design in biology does not reach that far."[26] Mathematician William Dembski likewise stressed that modern intelligent design theory is "more modest" than the old natural theology because it does not try to claim that science alone can get you to the God of the Bible. That is why "[i]ntelligent design as a scientific theory is distinct from a theological doctrine of creation. Creation presupposes a

Creator who originates the world and all its materials. Intelligent design attempts only to explain the arrangement of materials within an already given world. Design theorists argue that certain arrangements of matter, especially in biological systems, clearly signal a designing intelligence."[27]

Given that these comments from Dembski appeared in one of the two books advocating design that Peterson actually cites, it is mystifying how Peterson could get things so completely wrong. The mystery increases when later in his article Peterson seemingly contradicts himself and insists that "IDers will not say that the Intelligent Being behind nature is God."[28]

Well, which is it? Do "IDers" claim that intelligent design is "a hot, new scientific argument for God"? Or do they refuse to divulge whether the intelligent designer is God? The actual answer is neither. As the previous quote from Behe attests, intelligent design proponents who believe in God are certainly willing to say so. *What they are unwilling to do is claim that science alone can establish their belief in God.* They are unwilling to make this more expansive claim because they are being honest about the limits of modern science. Modern science cannot prove everything, and intelligent design theorists do not claim otherwise.

The third and perhaps most serious error in Peterson's description of intelligent design is his sweeping claim that the intelligent design movement "reject[s] evolutionary principles."[29] That is simply false, at least in the way that Peterson defines evolution.

Peterson's expansive definition of evolution encompasses not only "biological evolution" but "cosmic evolution," which he describes as "beginning with the Big Bang 13.7 billion years ago" and continuing with the production of galaxies and stars and planets.[30] Peterson's conflation of "cosmic evolution" and "biological evolution" into one meta-narrative is intellectually sloppy. "Cosmic evolution" and "biological evolution" are hardly the same thing. They are based on different evidence, make different claims, and raise different issues. Lumping them together may make the term "evolution" seem more grandiose, but it does almost nothing to illuminate the topic under discussion. Peterson's definition of "biological evolution" is just as unhelpful. It focuses entirely on the claim that biological life is the product of a long history of common descent without even identifying the mechanism that is supposed to make wholesale biological change possible.

The problem for Peterson is that modern intelligent design theory is not opposed to either "cosmic" or "biological" evolution as he defines those terms. Most intelligent design theorists certainly do not reject "cosmic evolution" beginning with the Big Bang. Indeed, many of the standard arguments for intelligent design in cosmology, physics, and astronomy are premised on the long history of the universe outlined by Peterson.[31] Nor is intelligent design incompatible with "biological evolution" when such evolution is defined merely as common descent. Although intelligent design theorists hold different views about the adequacy of the evidence for "universal" common descent (i.e., the idea that all organisms ultimately descend from the same common ancestor), they repeatedly make clear that intelligent design itself is compatible with such a belief. Thus, William Dembski writes that "[i]ntelligent design is compatible with both a single origin of life (i.e., common descent or monophyly) and multiple origins of life (i.e., polyphyly)."[32] Biochemist Michael Behe goes even further and affirms: "the idea of common descent (that all organisms share a common ancestor) fairly convincing, and have no particular reason to doubt it."[33] Discovery Institute, with whom most of the major proponents of intelligent design are affiliated, likewise states that if "evolution" is defined as the idea "that living things are related by common ancestry, then there is no inherent conflict between evolutionary theory and intelligent design theory."[34] Given that the quotations from William Dembski and Michael Behe just cited appeared in the only two books about intelligent design actually cited by Peterson, it is hard to understand how he could get this point wrong too. But he does.

Contrary to the muddied presentation in Peterson's article, the key tenet of "biological evolution" that intelligent design directly challenges is not common descent, but something Peterson curiously left out of his definition of evolution: The Darwinian claim that biological change is the product of an unguided process. Darwin was not the first person to propose that life had a long history or even to believe in common descent. His more important contribution was the hypothesis that such change over time could be achieved through a blind and undirected mechanism that could create fundamentally new things without the benefit of any foresight or planning. The mechanism that was supposed to achieve such wonders was natural selection acting on random variations (random mutations, according to modern scientists). It is important to understand that this claim that life is the product of a mindless and unguided process is not merely the invention of today's atheist Darwinists like Richard Dawkins. It was the core claim made by Darwin himself, and it is the core claim made by modern evolutionary theory today (sometimes known as "Neo-Darwinism").[35]

It is this in-your-face claim that biological change must be the product of an undirected process without foresight that intelligent design challenges based on the scientific evidence. Whether or not undirected material processes are capable of producing the exquisite beauty and functional complexity we see throughout the universe is *the* fundamental issue for intelligent design. Peterson's critique thus completely misses the point from the outset.

Peterson's confused view of intelligent design does not get any clearer later in his article. For example, he apparently thinks he is arguing against intelligent design when he claims that "Christians need not accept the notion that there are complex biological structures created directly by God without antecedent forms; they may hold a different view of how God brought about biological complexity."[36] But, again, intelligent design does not demand any such "notion." It does not rule out the existence of antecedent biological forms but only the Darwinian claim that such antecedents must have been produced as part of a mindless and undirected process.

Throughout his attack on intelligent design, Peterson seeks to convey the impression that intelligent design theorists reject the scientific method, reject "mainstream science," and are even "proposing God as a scientific explanation."[37] Wrong, wrong, and wrong again. Far from rejecting the scientific method, intelligent design theorists seek to apply the methods of modern science to the issue of detecting design in nature. Indeed, as Stephen Meyer meticulously argues in his book *Signature in the Cell*, modern intelligent design theorists employ the method of the historical sciences pioneered by Charles Lyell and adopted by Darwin himself. That method is based on the idea that "the present is the key to the past," and it seeks to explain past events by recourse to causes we see regularly operating today. As Meyer points out, we have abundant evidence from our own personal experience that intelligent causes can and do produce certain kinds of highly functional complexity (think computers, automobiles, and toasters). At the same time, we have abundant empirical evidence that non-intelligent causes do not seem able to produce these same kinds of highly functional complexity on their own today. Given this situation, the best explanation for the kinds of highly functional complexity under investigation is an intelligent cause.[38] Again, the logic of the design inference articulated by Meyer follows the very scientific method employed by Darwin to come to his conclusions. So if adopting this approach constitutes a rejection of the scientific method, Darwin himself stands guilty of the same charge.

As for the intimation that intelligent design theorists reject "mainstream science" in favor of "an alternative way of doing science," Peterson has confused raising questions about current scientific theories with rejecting "mainstream science." The present author knows quite a number of intelligent design theorists and scientists. He does not know of any who reject the periodic table, the germ theory of disease, the sphericity of the earth, the law of gravity, the discoveries of genetics, or any number of other findings of "mainstream science." Most of them embrace the standard models in cosmology and geology, and all of them accept core teachings of chemistry, cellular biology, mathematics, and similar disciplines. What they don't accept are dogmatic claims that unguided mechanisms are sufficient to explain the exquisite beauty and functional complexity we see throughout the biosphere and the universe. And when did questioning certain aspects of current scientific theories become a wholesale rejection of "mainstream science"? By that silly standard, Darwin himself was an opponent of "mainstream science," as are all scientists who propose new scientific theories or challenges old ones. Fortunately, "mainstream science" itself recognizes (at least in principle) that scientists need to be willing to reconsider old ideas based on new evidence. So instead of subverting mainstream science, intelligent design proponents are upholding its core commitment to open scientific inquiry.

Peterson's assertion that intelligent design scientists are "proposing God as a scientific explanation" is equally misplaced. Intelligent design theorists are proposing that science can detect the effects of intelligent causes within nature, not offering "God as a scientific explanation." There is a difference, and Peterson, as a philosopher, ought to be able to grasp it.

Peterson also offers up the hoary chestnut that intelligent design is based on "God of the gaps" reasoning that tries to fill current gaps in scientific knowledge with God. Wrong again. As Stephen Meyer points out, the modern version of the design inference is not an argument from ignorance, but an argument from knowledge. It based not on what we don't know about nature, but about what we *do* know. We have firsthand knowledge of what intelligent agents are capable of producing, and we have growing evidence about what unguided Darwinian natural selection *cannot* do in both the lab and the wild. Based on that extensive knowledge, design theorists argue that intelligent causation is the best explanation for certain features of nature.[39]

Ironically, it is Darwinism, not intelligent design, that has the real problem with reasoning from the "gaps." Darwinists typically embrace what biologist Jonathan Wells has aptly termed a "Darwin of the Gaps" approach.[40] Time and again, when functions for certain biological features have not been immediately apparent, Darwinists simply assume that the biological features must have been the product of a blind and undirected Darwinian process. This was the sort of flawed reasoning that led to the misclassification of the appendix and the tonsils as useless "vestigial organs," and which more recently inspired the colossal mistake of concluding that more than 90% of our DNA is "junk" because it doesn't code for proteins.[41]

A final point about intelligent design: While Peterson seeks to define "evolution" as encompassing everything from the formation of planets to the development of life, he tries to reduce modern intelligent design to mean irreducible complexity in biology, and then to suggest that this is completely different from "fine-tuning" arguments at the level of the universe. But this is an arbitrary distinction. Irreducible complexity is simply the "fine-tuning" argument applied to biology. Just as the laws of nature are finely tuned for the existence of life, many systems in biology seem to be exquisitely fine-tuned for their functions. These are conceptually the same kinds of arguments, which is why intelligent design theorists are interested in the fine-tuning of nature at all levels— from the universe as a whole to the operations of the cell to the chemical building blocks of life itself.

At this point, some readers are undoubtedly wondering what any of this has to do with C. S. Lewis's views on intelligent design. They are right to wonder: Although Michael Peterson's article is titled "C. S. Lewis on Evolution and Intelligent Design," in the end it has very little to do with Lewis. It is mostly a platform for Peterson to launch his own misguided critique of intelligent design. That is unfortunate. If Peterson had focused more on what Lewis said, he might have realized that Lewis has important lessons to teach us about the contemporary intelligent design debate. But they aren't the lessons he thinks, as we are about to see.

Lewis's Refutation of Seven Arguments Against Intelligent Design

Although we do not know how Lewis would have viewed the modern theory of intelligent design because he is no longer with us, we *do* know how Lewis would have responded to many of the standard arguments against intelligent design—because he responded to these same arguments when dealing with the issues of his own day. Lewis's responses to these arguments show just how skeptical he was of many standard materialist claims, and just how sympathetic he was to some of the main points raised by contemporary intelligent design theorists.

1. Unguided natural processes supersede the need for intelligent design.

ONE OF the most common arguments made today against intelligent design in biology is that intelligent design is unnecessary because we now *know* that complex biological features habitually emerge from simple parts through unguided evolutionary processes. Biologist Kathryn Applegate of the BioLogos Foundation, for example, claims that amazingly complicated molecular machines such as the bacterial flagellum (which functions like a high-tech outboard motor) self-assemble without any guiding intelligence because "[n]atural forces work 'like magic.'" Applegate continues: "It is tempting to think the spontaneous formation of so complex a machine is 'guided,' whether by a Mind or some 'life force' but we know that the bacterial flagellum, like countless other machines in the cell, assembles and functions automatically according to known natural laws. No intelligence required."[42]

One wonders whether Dr. Applegate draws the same conclusion every time she opens a spreadsheet program and discovers that it "magically" adds and subtracts sums—no intelligence required. Or when her word processing program "magically" checks the grammar and spelling of her blog posts—no intelligence required. One further wonders whether Dr. Applegate has ever visited a modern assembly line, where robotic equipment "magically" assembles any number of amazing products—no intelligence required. Of course, intelligence *is* required for each of these actions; the intelligence simply happens to be pre-programmed from the minds of men into the computer operations and assembly instructions. Similarly, the so-called magical assembly of the bacterial flagellum requires massive amounts of genetic information encoded in DNA, and as Stephen Meyer has persuasively argued, that information requires intelligence.[43]

As discussed earlier, Lewis thought that the sort of argument offered by Applegate was based on "a kind of optical illusion." Or to put it more strongly, "[t]he obviousness or naturalness which most people seem to find in the idea of emergent evolution... seems to be a pure hallucination."[44] Lewis observed dryly that in the real world "[w]e have never actually seen a pile of rubble turning itself into a house."[45] Instead, what we actually observe in nature are complex living things habitually arising out of equally complex living things, and simpler things that turn into complex things being preceded by the very complex things that they grow into. Acorns become oak trees not from something even simpler than an acorn, but from fully developed oak trees. Eggs that hatch into chickens ultimately arise not from undifferentiated protoplasm but from fully developed chickens who lay eggs. Molecular machines in bacteria ultimately arise not from simpler parts but from earlier bacteria that already have those same molecular machines. As a result, the ordinary physical processes of nature do not explain the actual origins of the complex functional features we find in nature; even less can they explain away the need for an intelligent cause for those features. "On any view, the first beginning must have been outside the ordinary processes of nature," wrote Lewis.[46] From our own experience of the creation of machines and human artifacts, the natural candidate for that outside cause is an intelligent designer according to Lewis.

2. Intelligent design is unnecessary because of the laws of nature.

A VARIATION of argument 1 is the claim that "natural laws" can create highly complex biological features without the need for intelligent design. Dr. Applegate alluded to this idea when she claimed that "the bacterial flagellum, like countless other machines in the cell, assembles and functions automatically according to known natural laws" and that this meant there was "[n]o intelligence required." Wrong again, according to Lewis, who pointed out that the "laws of nature" are absolutely incapable of causing anything on their own: "The laws of motion do not set billiard balls moving: they analyze the motion after something else (say, a man with a cue, or a lurch of the liner, or, perhaps, supernatural power) has provided it."[47] The laws of nature require input from outside, and if the effects caused are beyond the reach of blind chance ("a lurch of the liner") the input will need to come from an intelligent source ("a man with a cue" or a "supernatural power").

3. Intelligent design is a science-stopper.

ONE DOES not have to delve very deeply into current debates over intelligent design to encounter the claim that intelligent design is a "science stopper." But as Lewis made clear, it would be more correct to say that intelligent design is a science *starter*. "Men became scientific because they expected Law in Nature," wrote Lewis, "and they expected Law in Nature because they believed in a Legislator"—a.k.a. an intelligent designer. Thus, if people are concerned about the future progress of science, they should be worried about the abandonment of intelligent design by the scientific community: "In most modern scientists this belief [that behind nature is a Legislator] has died: it will be interesting to see how long their confidence in [the] uniformity [of nature] survives it... We may be living nearer than we suppose to the end of the Scientific Age."[48]

4. Intelligent design is simply an argument for God.

CRITICS TYPICALLY insist that modern intelligent design theory is simply an argument for God. However, as already explained, contemporary intelligent design theorists maintain that their version of the design argument is considerably more limited. In their view, evidence of design in nature may be enough to establish a purposeful cause for nature, but it does not answer all questions, such as the problem of evil, and so it cannot establish the existence of an all-wise, all-good, and all-powerful supernatural being taken by itself. Here is where Lewis's concern about the "argument from undesign" actually weighs in favor of the contemporary version of the design argument. Lewis essentially supports the more humble position of modern design theorists like Michael Behe and

William Dembski that evidence for design can refute materialism, but standing alone it is not enough to establish Christian theism.

5. Intelligent design is demeaning to God.

ALTHOUGH THE argument for intelligent design within science is not enough to establish the existence of God, it certainly has implications for those who already believe in God. If one happens to be a theist, it is natural to attribute the design of the world to God. But some theistic critics of intelligent design have taken to arguing that it is demeaning to God or nature to view God as a designer because then nature becomes somehow mechanical or God becomes merely an engineer. To those who make this argument, perhaps the best reply may be a single question asked by Lewis: "Would you make God less creative than Shakespeare or Dickens?"[49] God is certainly more than a designer. But do we dare contend that He is less? And the works of Shakespeare are certainly the products of intelligent design, but surely that does not make them mechanical or less beautiful.

6. Intelligent design is philosophy, not science.

ANOTHER ARGUMENT frequently employed to refute intelligent design is that it is "philosophy, not science." This argument is typically used to shut down conversations about the scientific evidence for intelligent design; but it also is typically applied inconsistently. Darwinian theory purports to provide scientific evidence that life is the product of an undirected process rather than intelligent design. Is *that* claim scientific? If it is, then so is intelligent design, because it purports to provide scientific evidence that bears on the very same question addressed by Darwinism—whether life is the product of a guided or unguided process. If it is scientific for supporters of Darwin's theory to offer empirical evidence and arguments *against* intelligent design, it should be equally scientific for supporters of intelligent design to offer empirical evidence and arguments in *favor* of intelligent design. Of course, perhaps Darwinian theory itself is philosophy rather than science, and then in that case perhaps intelligent design is too. But in either case, isn't the real issue determining what the truth actually is? Rather than trying to decide the debate over intelligent design by drawing arbitrary lines between science and philosophy (something notoriously difficult to do), why not focus on what evidence and logic actually show? At the basis of the "philosophy, not science" objection is the assumption that scientific and philosophical reasoning are two very different things that can never be mixed (there is usually an additional assumption as well on the part of scientists that scientific reasoning is superior to philosophical reasoning). Lewis provided a helpful corrective here, because he forcefully argued against the idea that scientific reasoning is substantially different (or better) from other kinds of reasoning. Contending that "the distinction... made between scientific and non-scientific thoughts will not easily bear the weight we are attempting to put on it,"[50] Lewis noted that "[t]he physical sciences... depend on the validity of logic just as much as metaphysics or mathematics." Thus, "[i]f popular thought feels 'science' to be different from all other kinds of knowledge because science is experimentally verifiable, popular thought is mistaken... We should therefore abandon the distinction between scientific and non-scientific thought. The proper distinction is between logical and non-logical thought."[51] Applied to the modern debate over intelligent design, Lewis's point means that the debate cannot be decided by drawing arbitrary lines between science and other disciplines.

7. Intelligent design is anti-science because it violates the scientific consensus.

INTELLIGENT DESIGN is frequently attacked as "anti-science," a charge that usually is based on no more than the bare fact that intelligent design proponents disagree with key parts of Darwinian theory. Since Darwinian theory is the "consensus view of science," challenging it makes one "anti-science." QED. The ridiculousness of this argument has been addressed already: If one follows the logic to its conclusion, Darwin himself would have to be declared anti-science for challenging the scientific consensus of his own day. So would Galileo. So would Einstein. As previously discussed in chapters 1, 4, and 6, Lewis provides an antidote to this kind of complaint by pointing out that the science of any given era may be driven more by larger cultural attitudes than the weight of the evidence. Lewis also had a keen appreciation for the radical changeability of science.[52] Thus, scientific beliefs cannot be regarded as sacrosanct, and those who challenge them should not be regarded as enemies of science any more than those who challenge at elections the existing party in control of government should be regarded as enemies of representative democracy. What is required in science is a robust exchange of ideas, not efforts to suppress legitimate debate. Hence, it is not an adequate refutation of intelligent design (or any other idea) to label it "anti-science" merely because it challenges the existing consensus.

Following the Argument Wherever It Leads

C. S. Lewis was a literary scholar, not a scientist, and so he did not feel it was his place to enter too deeply into the scientific debates of his own era. He also cautioned Christians about relying too heavily on the findings of science for their apologetics. After all, the findings of science are in a constant state of flux. At the same time, as C. John Collins explained in chapter 5, Lewis was willing to draw on the insights of science in his own apologetics. Perhaps more importantly, Lewis urged Christians with a scientific aptitude to keep up with the science of their day because "[w]e have to answer the current scientific attitude towards Christianity, not the attitude which scientists adopted one hundred years ago." He further encouraged Christians to write books about science that would counter the materialist worldview implicitly by presenting "perfectly honest" science.[53]

Most important of all, Lewis was a consistent champion of following an argument wherever it might lead, without placing artificial barriers to the consideration of new ideas. Many people do not realize just how much Lewis modeled this principle in his own life, or how he encouraged students to adopt the credo as their own. A good example is his role in founding the Socratic Club at Oxford University. From 1942 until he left for Cambridge University in 1954, Lewis served as President of the club, a weekly gathering of students and scholars devoted to living out the injunction of Socrates to "follow the argument wherever it led them," especially in debating the truth or falsity of Christianity.[54] "We never claimed to be impartial," remembered Lewis. "But argument is. It has a life of its own. No man can tell where it will go."[55]

The Oxford Socratic Club likely had a profound effect on many students, but no more so than on one regular attendee who later recalled that "the Socratic principle I saw exemplified there—of following the evidence wherever it may lead—increasingly became a guiding principle in the development, refinement, and sometimes reversal of my own philosophical views."[56]

The attendee in question was a young Antony Flew, who went on to play an important role in legitimizing the contemporary debate over intelligent design. Eventually becoming one of the most noted atheist philosophers in academia, Flew startled the world in 2004 by publicly renouncing his atheism in favor of a belief in God (although not Christianity).[57] Following the credo he had seen embodied by Lewis's Socratic Club, Flew had continued to follow the evidence until it led him to a complete change of mind.

Flew credited new scientific evidence for intelligent design as a key reason for his conversion. As he told one interviewer in 2004, "I think the argument to Intelligent Design is enormously stronger than it was when I first met it."[58] Flew's reading had included books by intelligent design theorists Michael Behe and William Dembski, and he was especially influenced by the argument for design based on the biological information encoded in DNA.[59]

In the end, Lewis's greatest contribution to the intelligent design debate may have been his steadfast insistence to Flew—and many others—that they should pursue an argument wherever it might lead. That insistence inspired Flew to consider seriously new evidence for intelligent design despite the prejudices of the existing intellectual establishment. And the evidence changed his mind.

As Lewis said, an argument "has a life of its own. No man can tell where it will go."

Endnotes

1. C. S. Lewis, *Spirits in Bondage: A Cycle of Lyrics*, edited by Walter Hooper (San Diego: Harcourt Brace & Co., 1984); also available online at http://www.gutenberg.org/ebooks/2003, accessed June 23, 2012.
2. "I: Satan Speaks," ibid., 3.
3. C. S. Lewis, *Surprised by Joy* (New York: Harcourt Brace Jovanovich, 1955), 65.
4. Ibid.
5. Pascal, quoted in C. S. Lewis, *The Problem of Pain* (New York: Macmillan, 1962), 13.
6. Lewis to Dom Bede Griffiths, Dec. 20, 1946, in *The Collected Letters of C. S. Lewis* (San Francisco: HarperSanFrancisco, 2004), vol., II, 747.
7. Lewis, *Surprised by Joy*, 65.
8. The phrase is Tennyson's, but one can hear echoes of it in Lewis's poem "Ode for New Year's Day": "For Nature will not pity, nor the red God lend an ear." C. S. Lewis, "VIII: Ode for New Year's Day," *Spirits in Bondage*, 14. For Tennyson's line see "In Memoriam A. H. H." (1850), Canto 56, accessed June 25, 2012, http://www.online-literature.com/donne/718/.
9. See Lewis, *Surprised by Joy*, 7, 16, 152–158. What I am describing here as the argument from natural beauty is one part of Lewis's larger argument from "joy." For a helpful overview of the development of this argument, see Kathryn Lindskoog, *Finding the Landlord: A Guidebook to C. S. Lewis's Pilgrim's Regress* (Chicago: Cornerstone Press, 1995), xiii–xxv; also see Lewis, "Surprised by Joy;" Lewis, "The Weight of Glory," in C. S. Lewis, *The Weight of Glory and Other Addresses*, revised and expanded edition, edited by Walter Hooper (New York: Macmillan, 1980), 3–19.
10. "XXVI: Song," Lewis, *Spirits in Bondage*, 50.
11. See Lewis to Dom Bede Griffiths; C. S. Lewis, "De Futilitate," in *Christian Reflections*, edited by Walter Hooper (Grand Rapids, MI: Eerdmans, 1967), 65–67, 69–70.
12. See C. S. Lewis, *Miracles: A Preliminary Study*, 1960 edition (New York: Macmillan, 1978), especially, 12–24; Lewis, "De Futilitate," 57–71; Lewis, "Religion without Dogma?" in *God in the Dock*, edited by Walter Hooper (Grand Rapids: Eerdmans, 1970), 129–146.
13. C. S. Lewis, "Is Theology Poetry?" in *The Weight of Glory and Other Addresses*, revised and expanded edition (New York: Macmillan, 1980), 90; also see C. S. Lewis, "Modern Man and His Categories of Thought," *Present Concerns*, edited by Walter Hooper (New York: Harcourt Brace Jovanovich, 1986), 63–64.
14. Lewis, "Is Theology Poetry?," 90.
15. Lewis, "Two Lectures," *God in the Dock*, 211.
16. Lewis, "Is Theology Poetry?," 91.
17. Ibid.
18. Michael L. Peterson, "C. S. Lewis on Evolution and Intelligent Design," *Perspectives on Science and the Christian Faith* 62, no. 4 (December 2010): 253–266; Michael L. Peter-

son, "C. S. Lewis on Evolution and Intelligent Design," www.biologos.org (April 2011), accessed May 18, 2012, http://biologos.org/uploads/projects/peterson_scholarly_essay.pdf.
19. Peterson, "C. S. Lewis on Evolution and Intelligent Design" (2010), 255.
20. C. S. Lewis, *The Voyage of the Dawn Treader* (New York: Macmillan, 1952), 133.
21. Michael Behe, *Darwin's Black Box: The Biochemical Challenge to Evolution* (New York: Free Press, 1996); William Dembski, *Intelligent Design: The Bridge Between Science and Theology* (Downers Grove, IL: InterVarsity, 1999). In contrast to Peterson's outdated citations to books and articles by intelligent design proponents, he cites several books from recent years attacking intelligent design. See Peterson, "C. S. Lewis on Evolution and Intelligent Design" (2010), note 13, 265.
22. Stephen C. Meyer, *Signature in the Cell: DNA and the Evidence for Intelligent Design* (New York: HarperOne, 2009). In the BioLogos version of Peterson's article, the inaccurate reference to Meyer's book has been corrected.
23. Peterson, "C. S. Lewis on Evolution and Intelligent Design" (2010), 254.
24. James E. Horigan, *Chance or Design?* (New York: Philosophical Library, 1979); Charles B. Thaxton, Walter L. Bradley, and Roger L. Olsen, *The Mystery of Life's Origin* (Dallas: Lewis and Stanley, 1992 [second printing]); Michael Denton, *Evolution: A Theory in Crisis* (Bethesda, MD: Adler and Adler, 1985).
25. See Jonathan Witt, "A Brief History of the Scientific Theory of Intelligent Design," accessed June 23, 2012, http://www.discovery.org/a/3207.
26. Michael Behe, "The Modern Intelligent Design Hypothesis," *Philosophia Christi*, 3, series 2, no. 1 (2001): 165.
27. Dembski, *Intelligent Design: The Bridge Between Science and Theology*, 248.
28. Peterson, "C. S. Lewis on Evolution and Intelligent Design" (2010), 256.
29. Ibid., 254.
30. Ibid.
31. For discussion of these arguments by scholars supporting intelligent design, see Guillermo Gonzalez and Jay Richards, *The Privileged Planet: How Our Place in the Cosmos is Designed for Discovery* (Washington, D.C.: Regnery, 2004), and Stephen Meyer, "A Scientific History and Philosophical Defense of the Theory of Intelligent Design" (2008), accessed June 25, 2012, http://www.discovery.org/a/7471.
32. Dembski, *Intelligent Design: The Bridge Between Science and Theology*, 250.
33. Behe, *Darwin's Black Box*, 5.
34. Answer to the question "Is intelligent design theory incompatible with evolution?," *Top Questions*, accessed June 25, 2012, http://www.discovery.org/csc/topQuestions.php.
35. For documentation of the fact that Darwinian evolution is supposed to be unguided and lacking in foresight, see discussion in chapter 6 in addition to West, *Darwin's Conservatives: The Misguided Quest* (Seattle: Discovery Institute Press, 2006), 13–17; and West, "Nothing New Under the Sun: Theistic Evolution, the Early Church, and the Return of Gnosticism, Part 1," in Jay Richards, editor, *God and Evolution: Protestants, Catholics, and Jews Explore Darwin's Challenge to Faith* (Seattle: Discovery Institute Press, 2010), 38–40.

36. Peterson, "C. S. Lewis on Evolution and Intelligent Design" (2010), 259.
37. Peterson, "C. S. Lewis on Evolution and Intelligent Design" (2010), 257, 259.
38. Meyer, *Signature in the Cell*, especially 150–172, 324–348; and Meyer, "A Scientific History," 11–17.
39. Meyer, "A Scientific History," 27–28; also see Wells, "Darwin of the Gaps," *God and Evolution*, 121–122.
40. Wells, "Darwin of the Gaps," 128.
41. See Casey Luskin, "Vestigial Arguments about Vestigial Organs Appear in Proposed Texas Teaching Materials," *Evolution News and Views*, June 20, 2011, accessed May 19, 2012, http://www.evolutionnews.org/2011/06/vestigial_arguments_about_vest047341.html; David Klinghoffer, "Looks Like the Appendix Isn't a 'Junk Body Part' After All," *Evolution News and Views*, Jan. 4, 2012, accessed May 19, 2012, http://www.evolutionnews.org/2012/01/now_its_the_app054761.html; Jonathan Wells, *The Myth of Junk DNA* (Seattle: Discovery Institute Press, 2011).
42. Kathryn Applegate, "Self-Assembly of the Bacterial Flagellum: No Intelligence Required," *The BioLogos Forum*, August 19, 2010, accessed June 23, 2012, http://biologos.org/blog/self-assembly-of-the-bacterial-flagellum-no-intelligence-required/.
43. Meyer, *Signature in the Cell*.
44. Lewis, "Is Theology Poetry?," 90–91.
45. Lewis, "The Funeral of a Great Myth," *Christian Reflections*, 90.
46. Lewis, "Two Lectures," 211.
47. Lewis, *Miracles*, 58–59.
48. Ibid., 106.
49. Ibid., 65.
50. Lewis, "De Futilitate," 61.
51. Ibid., 62.
52. Lewis, "Christian Apologetics," *God in the Dock*, 92; Lewis, *The Discarded Image* (Cambridge: Cambridge University Press, 1964), 219–223.
53. Lewis, "Christian Apologetics," 93.
54. "The Founding of the Oxford Socratic Club," *God in the Dock*, 126.
55. Ibid., 128.
56. Antony Flew with Roy Abraham Varghese, *There Is a God* (New York: HarperOne, 2007), 42.
57. "Lifelong Atheist Changes Mind about Divine Creator," *The Washington Times*, Dec. 9, 2004, accessed June 25, 2012, http://www.washingtontimes.com/news/2004/dec/9/20041209-113212-2782r/?page=all.
58. Antony Flew interview with Gary Habermas for the journal *Philosophia Christi* (Winter 2004), accessed June 25, 2012, http://www.biola.edu/antonyflew/page2.cfm.
59. Gene Edward Veith, "Flew the Coop," *World*, Dec. 25, 2004, accessed June 25, 2012, http://www.worldmag.com/articles/10094. In addition to Behe and Dembski, Flew was heavily influenced by intelligent-design-friendly physicist Gerald Schroeder. On the design implications of DNA, he cited an article by Discovery Institute Senior Fel-

low David Berlinski. See Flew, *There Is a God*, 127. It should be noted that Michael Peterson in his article wrongly implies that Flew's conversion was either not connected to intelligent design or even hostile to it. Michael Peterson, "C. S. Lewis on Evolution and Intelligent Design" (2010), 255, 259.

REASON

Unless human reasoning is valid, no science can be true.

—C. S. Lewis, Miracles

8

MASTERING THE VERNACULAR

C. S. LEWIS'S ARGUMENT FROM REASON

Jay W. Richards

C. S. LEWIS SAVED MY LIFE—MY INTELLECTUAL LIFE AT LEAST. As a college freshman I had vaguely but poorly developed Christian beliefs, and found myself in an environment that seemed designed to corrode them. One required course exposed students to reports of peculiar foreign cultures and religious mythologies. The purpose? To persuade us, the students, that our beliefs were merely residue from our particular time and place, with more no claim to truth than the exotic cosmologies of the native New Guineans. The course, "Freshman Symposium," should have been titled "How to Turn a Freshman into a Relativist in Twelve Weeks." Though a relativist fog seemed to linger over most of my classes, Freshman Symposium was clearly meant to secure it in the minds of impressionable eighteen-year-olds.

In addition to relativism, another idea, which I later learned to call naturalism or materialism, was everywhere assumed but never quite defended. The closest thing to an argument for naturalism that I encountered was along these lines: Primitive peoples, including ancient Jews and Christians, believed in all manner of spiritual beings that could act in the world. They supposed that angels could give messages to the faithful and that God could cause a virgin to become pregnant, raise Jesus from the dead, turn water into wine, and so forth. In our day and age, however, what with telephones and radios and microwave ovens, we can no longer believe such nonsense.

As a philosophical argument, this isn't impressive. But it still had *sociological* force for a defenseless college student. The last thing you want to be, as a college freshman, is intellectually unfashionable. If all the smart people around you believe that the moon is made of green cheese, you may find yourself entertaining the green cheese hypothesis, even if you've never encountered any evidence for it.

I detected this naturalistic assumption in my science and economics courses, but *especially* in religious and biblical studies. For instance, I learned that New Testament books, such as the gospels, had been written several generations after the events they supposedly reported. After all, the gospels seemed to describe Jesus predicting the destruction of the Temple. The Temple was destroyed in 70 AD, so the gospels *had to be* written after that date. Moreover, the gospels described Jesus's resurrection, so they must have been written late enough for a resurrection legend to have time to develop. The unstated missing premise was easy enough to detect: "We now know" that predictions of future events and resurrections are impossible.

At first encounter, I really didn't know what to think about these arguments, and found myself starting down the path of Protestant liberalism, in which all the supernatural doctrines of Christianity are reinterpreted or "demythologized" in more fashionable terms. The incarnation, for instance, is reinterpreted to mean that people are special, and the resurrection means, not that Jesus was raised from the dead, but that we should have hope in the future. This strategy was very unattractive, however, and before long, I decided that this way of resolving difficulties was a spineless cop-out.

I would probably have become an agnostic. But someone (I don't know who) had given me a little "six packet," called Signature Classics, of C. S. Lewis books: *Mere Christianity, The Screwtape Letters, The Abolition of Man, The Problem of Pain, Miracles,* and *The Great Divorce*.[1] As it happened, one of the books, *The Abolition of Man*, provided an excellent refutation of relativism, and another, *Miracles*, exposed the problems with naturalism. Though Lewis wrote these books in the 1940s, their basic arguments still applied to the American college scene four decades later.

When it came to relativism, I already realized that just because people believed all sorts of different things, it didn't follow that they were all wrong. Moreover, I sensed the self-refuting quality of relativism almost immediately—a truth claim that there are no truths is just nonsense. But Lewis's *Abolition of Man* gave me confidence in my vague intuitions.

Still, I realized that to go on believing that God existed and acted in the world, I needed more than merely the fact that I had grown up believing such things. I needed serious, positive arguments against naturalism and for Christian theism. *Miracles* introduced me to such arguments.

The Consummate Translator

I OFFER THESE AUTOBIOGRAPHICAL DETAILS because I think they reveal something of Lewis's genius. He was an accomplished academic with specialties in literary criticism and medieval literature. But he also was conversant with the broader debates then current in the academy. He had what we would now call a "photographic memory." He knew a great deal about the history of ideas, and how the ideas of his day fit within that history. Although idealism was fashionable when he was a student, by the time he started writing apologetics, idealism had been displaced by positivism. And yet he was able to transcend these intellectual orthodoxies and critique them persuasively for the *ordinary reader* who lacked such background.

When I first read *Miracles*, I was a confused eighteen-year-old who had never had a philosophy course and didn't know what it meant to beg a question or argue in a circle. And yet I found Lewis's arguments both accessible and persuasive. Perhaps more astonishing, I have now spent years studying and engaging philosophy and Christian apologetics, and I still find Lewis's arguments, for the most part, persuasive.

C. S. Lewis was the consummate translator. This is an academic achievement every bit as impressive and lasting as any other. Translation of academic subjects into laymen's terms is akin to hand-copying Van Gogh's *Starry Night*, but with a much more limited palette of colors than the great Dutch artist used for the original. The original required artistic genius. But a good copy using a limited palette requires genius as well.

Lewis understood the difficulty of good translation. He once observed, "Any fool can write *learned* language. The vernacular is the real test."[2] Many academics, in contrast, disdain the task of translation. They seem to pride themselves on grinding out turgid academic prose that is accessible to few and enjoyable to none. They write largely for those within their narrow academic discipline. Lewis never settled for such a provincial academic career. On the contrary, he made his own academic life difficult by writing children's books and Christian apologetics. Most Lewis scholars suspect that this is one reason he never advanced beyond the title of lecturer during all his years at Oxford University. It was only late in life that Cambridge University had the good sense to hire him and give him a professorial title befitting his academic stature.

To call Lewis a popularizer seems demeaning. For lack of a better term, we might say that Lewis was an "elite populist" or "elite generalist" who dwelled comfortably in two very different worlds. We must distinguish the elite populist from the dabblers or "second hand peddlers of ideas"—in Ludwig von Mises's biting words—who have a disproportionate but mostly undeserved influence on culture. Such pundits offer their opinions on everything from film criticism and science to economics and politics; but their commentary is often superficial because they haven't first learned those subjects. Rather than translating, they merely opine.

Still less is an elite populist such as Lewis akin to the person who has expertise in one area, then takes up another area late in life, and imagines himself to have made some groundbreaking discovery. Such a person is guilty of fake originality, because he has not read widely or deeply enough in the field to realize that he is walking on well-travelled soil.

The Argument from Reason

To see Lewis's genius as an elite populist or translator, I'd like to focus on one of his best-known arguments—often called the "argument from reason." The purpose of the argument is to show that naturalism and reason are incompatible, that believing in naturalism is self-defeating. That is, if naturalism is true, then we ought not to trust our capacity for reason, and so, ought not to trust arguments in favor of naturalism.

Philosopher Victor Reppert describes the argument (and several versions he develops from Lewis's original) as "beginning with the insistence that certain things must be true of us as human beings in order to ensure the soundness of the kinds of claims we make on behalf of our reasoning."[3] This argument gained attention when Lewis proposed it in the first edition of *Miracles*. Philosopher Elizabeth Anscombe critiqued the original formulation of the argument, so Lewis corrected it in a subsequent edition of *Miracles*.[4] It is this revised version of his argument that millions of readers have encountered. (He also discusses the argument in some lesser-known articles published in *Christian Reflections* and *God in the Dock*.)

Lewis taught philosophy in his first year as a lecturer at Oxford, but he wasn't a professional philosopher. Moreover, he made the argument from reason in a small book written for public consumption. And yet it has been remarkably resilient and fruitful. Its philosophical offspring still play a role in contemporary philosophy. The most rigorous form of the argument is the "Evolutionary Argument Against Naturalism" developed and refined by philosopher Alvin Plantinga. We will consider Plantinga's argument, but first let's look at Lewis's formulation.

The Basic Argument

Miracles is not a historical defense that miracles such as the Incarnation and Resurrection have actually occurred. It is a preliminary defense of their possibility and propriety. One of its central arguments is that we cannot determine the antecedent probability of a miracle without first deciding what reality is like. If you think that a transcendent

God exists, for instance, you will assess the evidence for a miracle differently than if you are a naturalist who believes that the closed, interlocking system of nature is all that exists. As a result, Lewis spends a good deal of time in *Miracles* evaluating the competing claims of what he calls Supernaturalism and Naturalism.

It is in this context that Lewis takes up the so-called "cardinal difficulty of naturalism." Naturalists in Lewis's day were very much like naturalists in our day. They normally imagine that their philosophy is the result of sound reasoning and solid evidence, and assume non-naturalists are ignorant and irrational. Lewis argues quite the opposite: Naturalism is not compatible with knowledge and the reliability of reason.

Naturalists, like everyone else, generally trust their reason to lead them to truth. We all take it for granted that we can learn about the world around us through our senses. We experience heat and sound and color and other people. We somehow synthesize and take account of these things with our mind. From these experiences we make inferences about the world: "We infer Evolution from fossils: we infer the existence of our own brains from what we find inside the skulls of other creatures like ourselves in the dissecting room."[5]

But what is an inference? Clearly it's not an object of the senses, such as a bullfrog or a forest fire. An inference, we might say, is a logical structure. We see dinosaur fossils with our eyes; but we infer the prior existence of dinosaurs with our minds. We simply understand that if all men are mortal and Plato is a man, then Plato is mortal. When we consider this argument, we're not observing the world around us. *We're perceiving logical relations between propositions* that would obtain in any possible world. Propositions are claims about states of affairs, but not the states of affairs themselves. *That Columbus sailed the ocean blue* is a state of affairs that obtained sometime in 1492. "In 1492, Columbus sailed the ocean blue" is an English formulation of a true proposition affirming that state of affairs.

When we conclude that Plato is mortal, we are certain that if the propositions that form the premises are true, then the conclusion *must* be true. Unless such mundane inferences are possible and reliable, we can't have knowledge:

> All possible knowledge... depends on the validity of reasoning. If the feeling of certainty which we express by words like *must be* and *therefore* and *since* is a real perception of how things outside our own minds really "must" be, well and good. But if this certainty is merely a feeling *in* our own minds and not a genuine insight into realities beyond them—if it merely represents the way our minds happen to work—then we can have no knowledge. Unless human reasoning is valid no science can be true.[6]

Naturalism doesn't contain such ingredients as minds, propositions, perceptions and logical relations. It contains elementary particles and attractive forces, chemical reactions, quantum fields, and the like, in a closed and impersonal system of cause and effect. And all of those causes are material and non-rational. Naturalism doesn't countenance immaterial entities such as persons, with thoughts and beliefs, persons that can infer from the proper ground of a propositional belief to a valid conclusion, which can then guide behavior and so cause things to happen in the world. If naturalism is true, then all these "things" either don't exist or must have some non-rational *physical* cause. And we have no reason to think that such causes would provide us with a way of inferring correctly from a ground to a consequent (as Lewis puts it).

But naturalists normally trust the conclusions of natural science and typically believe they have arrived at their naturalist convictions by following evidence and sweet reason to their inevitable conclusion. If Lewis is right, however, then naturalism as a belief refutes itself. Consider any argument for naturalism. If it is sincerely offered, it will presuppose that people have beliefs and rational faculties that can affect their action because they can perceive the validity of the argument, or lack thereof, and act accordingly.

Even prominent naturalists have admitted this dilemma. Lewis quotes famous naturalist J. B. S. Haldane to this effect. "If my mental processes are determined wholly by the motions of atoms in my brain," said Haldane,"I have no reason to suppose that my beliefs are true… and hence I have no reason for supposing my brain to be composed of atoms."[7] (Haldane's reference is to materialism—the idea that the fundamental constituents of reality are bits of matter. One could be a naturalist but believe nature consists of something other than matter. For our purposes, however, we'll treat materialism and naturalism as synonyms.)

In fact, even Darwin himself admitted the worry:

> With me the horrid doubt always arises whether the convictions of man's mind, which has been developed from the mind of the lower animals, are of any value or at all trustworthy. Would any one trust in the convictions of a monkey's mind, if there are any convictions in such a mind?[8]

The "cardinal difficulty of naturalism" doesn't depend on a debatable theistic assumption. It emerges from the lack of causal tools in the naturalist toolkit. Strictly speaking, Lewis's argument doesn't show that naturalism is false, so much as it shows that naturalism can't be rationally believed. Again, if naturalism were true, then beliefs, purposes, and inferences either wouldn't exist or wouldn't have any obvious power to transmit truth and so wouldn't give us real knowledge of the world. That would apply to naturalistic belief as well. If naturalism were true and we believed in naturalism, we would lack the rational, truth-conducive faculties to consistently believe that it's true, even less to know it.

If, in contrast, the fundamental reality is mind or reason—if everything proceeds from the wisdom and counsel of an eternal God, if human beings are made in the image of God to know and seek truth—then we would expect that our reasoning is at least sometimes reliable with respect to those matters that our reason was designed to grasp.

The naturalist who is unfamiliar with this argument invariably invokes Darwin at this point. Naturalists tend to believe that Darwin's account of the evolution of life is roughly correct. And they think the evidence establishes it. According to the Darwinian story, the adaptations of living things to their environment are *not* the result of purposeful design by God or an intelligent agent, but are the result of a blind process of natural selection acting on random variations within a population. Natural selection preserves and then propagates those variations that provide organisms with a survival advantage, and weeds out those that don't. While there are other factors in evolution—genetic drift, bottle-necking, and so forth—this process of selection-and-random-variation largely creates these adaptations, according to Darwinian theory. This process is not mere chance or randomness; but it is blind and unconscious. There is no agent choosing variations, such as genetic mutations, based on the survival advantage that they confer on an organism, or for any other reason.

If this story is roughly correct, then there would seem to be a survival advantage in forming true beliefs. Surely our ancestors would have gotten on in the world much better if they came to believe that, say, a saber-tooth tiger, is a dangerous predator. And if they believed that they should run away from dangerous predators, all the better. In contrast, those early humans who had false beliefs, who believed that saber-tooth tigers were really genies who would give three wishes if they were petted, would tend to get weeded out of the gene pool. So wouldn't the Darwinian process select for reliable rational faculties, and so, give us faculties that would produce true beliefs?

Lewis argues that this process—which preserves survival-enhancing features—is nevertheless non-rational, and so cannot be expected to produce rational faculties. Again, if naturalism were true, then one would not expect minds and agents, choices and intentions to exist at all. If these things did exist, surely they would be mere epiphenomena of physical states. But let's grant their existence, and even allow the naturalist the luxury of assuming that beliefs can guide our behavior. The naturalist will then want to argue that our reason and belief-forming faculties have been shaped by natural selection over eons, and so should be quite reliable.

The problem is that there are millions of beliefs, few of which are true in the sense that they correspond with reality, but all compatible with the same behavior. Natural selection could conceivably select for survival-enhancing *behavior*. But it has no tool for selecting only the behaviors caused by true beliefs, and weeding out all the others. So if our reasoning faculties came about as most naturalists assume they have, then we have little reason to assume they are reliable in the sense of giving us true beliefs. And that applies to our belief that naturalism is true.

This argument wasn't original with Lewis. It appears in the Gifford Lectures given by British statesman Arthur Balfour in 1914. At the time, Balfour's lectures were well-known. They were even reported individually in the newspaper, and eventually published as the book *Theism and Humanism*, which Lewis credits as one of the ten books that most influenced him.[9] But it is Lewis's form of the argument that is still published, and read, in the twenty-first century.

The Accessible Midpoint between the Metaphysical and Analytic

MUCH MORE CAN BE SAID in defense of this argument. Since Victor Reppert does just that in his chapter, however, I want to do something different: to locate it rhetorically. As it happens, Lewis's argument occupies the middle point between a general metaphysical argument common in traditional Western philosophy and a tight but technical argument from contemporary analytic philosophy.

The reader who knows the history of Western philosophy may recognize that Lewis's argument is a form of a traditional metaphysical argument, namely, that *an effect cannot be greater than its cause*, or, to put it otherwise, a cause cannot give a greater perfection than it has itself. Lewis's argument says that we cannot get a rational effect from a non-rational cause.

This metaphysical argument often functioned more as a self-evident principle than as an argument. Let's call this the "causal principle." For apologetics, the causal principle can be quite handy, since it makes it easy to show that there must be an uncaused and so eternal First Cause, a necessarily existing cause greater than and so necessary to cause everything else, including every other causal entity. And a necessarily existing, uncaused First Cause is what all men call God (as Thomas Aquinas put it). Accept the principle and you get a First Cause as a bonus.

If you have been trained in traditional Western metaphysics, the causal principle can seem not only true, but also intuitively obvious. Unfortunately, we live in an anti-metaphysical age, so an argument that employs the causal principle probably won't persuade anyone who does not already believe it.

I once saw a philosopher use this principle to prove the existence of God. Most of the well-educated audience, however, found the argument quite obscure. They didn't seem to understand what it would mean for an effect to be "greater" than its cause, or a cause to be greater than its effect. The whole idea of perfection seemed baffling to them. Worse, they had the impression that the causal principle, even if intelligible, was demonstrably false. After all, cosmic and biological evolution seem to provide examples of simple causes—physical constants, molecules, natural selection—giving rise to much more interesting and complex effects, such as stars, planets, and people. So the philosopher's argument was unpersuasive to the target audience.

Even if you believe the causal principle is valid—as I do—it seems to compress most of the hard work you expect to get from argument and evidence into a single premise. Worse, in most contemporary contexts, it doesn't serve as a useful premise in an argument because it requires an intellectual apparatus that most people either lack or doubt altogether.[10]

I have no doubt that Lewis knew and believed the causal principle. Because he was a consummate translator, however, he didn't use it in its traditional form. He knew intuitively that a good, persuasive argument is not only sound, but also appeals to something the target audience already knows or believes. Rather than invoking an abstract metaphysical principle, then, Lewis appealed to two beliefs already held by his target audience—the validity of reasoning and the truth of naturalism. Lewis argued that one could not consistently believe both.

Plantinga's Evolutionary Argument Against Naturalism

Lewis's achievement was to make the argument from reason precise enough to be persuasive but accessible enough to be read and understood by millions. This is an important intellectual achievement. Still, some stout souls seek a really rigorous, airtight version of Lewis's argument. For them, philosopher Alvin Plantinga has delivered.

Plantinga first presented his "evolutionary argument against naturalism" (EAAN) in 1991, but he has continued to refine it in several publications since then. He offers what he hopes is the "official and final" version in his insightful 2011 book *Where the Conflict Really Lies: Science, Religion, & Naturalism*.[11]

While Lewis focused on our reason, Plantinga focuses on our cognitive faculties in general, "those faculties, or powers, or processes that produce beliefs or knowledge in us."[12] These include memory, perception, intuitions about logical and mathematical truths, and perhaps other faculties, such as our ability to discern the thoughts and feelings of others, and to know of moral truths and the existence of God. The EAAN concerns these faculties, which all of us assume are generally reliable over the range of ordinary subjects. For instance, we assume that our memory is generally reliable, that we can know certain logical and mathematical truths and perceive the world around us.

The first premise of the argument is what Plantinga calls "Darwin's doubt." This is the idea that Darwinian evolution seems inadequate to produce generally reliable cognitive faculties. The doubt is expressed, not just by Darwin and J. B. S. Haldane, as we saw above, but also by naturalists and atheists such as Nietzsche, Thomas Nagel, Barry Stroud, and Patricia Churchland. Churchland's formulation is surely the most colorful:

> Boiled down to essentials, a nervous system enables the organism to succeed in the four F's: feeding, fleeing, fighting and reproducing. The principle chore of nervous systems is to get the body parts where they should be in order that the organism may survive... Improvements in

sensorimotor control confer an evolutionary advantage: a fancier style of representing is *advantageous so long as it is geared to the organism's way of life and enhances the organism's chances of survival*. Truth, whatever that is, definitely takes the hindmost.[13]

Darwinian evolution, in other words, selects at most for survival-enhancing behavior, not true beliefs.

Plantinga sees that this isn't a deductive argument. It's logically possible that we could have reliable cognitive faculties. So he frames Darwin's doubt in terms of conditional probability. Given naturalism (N) and the assumption that we are the product of Darwinian evolution (E), the probability (P) that we have reliable cognitive faculties (R) is low. It's logically possible but unlikely. The premise can be abbreviated in logical notation as:

P(R/N&E) is low.

Most naturalists are also materialists in the sense that they think human beings are just material objects, full stop. No mind, agent, or person exists that is distinct from the body or the brain. The very idea of reliable cognitive faculties—beliefs, reason, and the like—is hard to accommodate in this framework. The only tool in the naturalist toolkit with any hope of giving us such faculties would be Darwinian evolution.

So given materialism and Darwinian evolution, beliefs will have to be some kind of event or structure in the brain. They will be "*electrochemical* or *neurophysical properties* (NP, for short)."[14] If these are beliefs, however, NP will also have propositional content, such as: *If A, then B. A, therefore, B*, or *That lion looks hungry*.

The reductive materialist will say that the propositional content of a belief is identical with some neurophysical property, perhaps some cluster of neurons somewhere in the brain. (Frankly, I don't think this claim is even intelligible, but it is what the reductive materialist will say.) The nonreductive materialist will say that the propositional content of a belief can't be reduced to a neurophysical property; but that the content is determined by NP. In both accounts, "mental properties are determined by physical properties."[15]

NP may have physical effects. That is, some neurophysical property may cause a person to run from predators, and so have selective advantage. But the propositional content—the truth or falsity of the belief that either is or corresponds to that NP—need not be relevant to the causal connection between an NP and its behavioral effect. The behavioral effect and perhaps the neurophysical effect that causes it would be selectable by natural selection. But assuming naturalism, natural selection and random genetic mutation would be blind to the propositional content that either is or is determined by the NP.

Just think about how we are supposed to have arisen according to the naturalist and Darwinian story. By some unknown combination of chemistry and dumb luck, a reproducing organism emerged on the early earth. For several billion years, single-celled bacterial organisms populated the earth. Eventually more complex, multicellar organisms emerged. During all that time, the blind processes of natural selection and random genetic mutations were preserving organisms in populations with slight survival advantages. Surely such organisms lacked beliefs and consciousness altogether. About half a billion years ago, many animal phyla arrived on the scene, and somewhere along the line, beliefs began to spin off from certain neurophysical structures or events that are either caused by or correlated with certain survival-enhancing behaviors.

To take ourselves out of the scenario, Plantinga suggests that we imagine a world distinct from our own that is very much like the one we described above. In this world, naturalism is true and no God exists. At some point, very late in the game, creatures similar to us begin to have beliefs. Now, given this scenario, how likely is it that their beliefs, and their cognitive faculties, will be reliable, in the sense that they generally produce beliefs with true propositional content? The obvious answer is, "Not very likely."

Plantinga gives more meat to the first premise of the evolutionary argument against naturalism than we can summarize here, but the broad outlines should be clear enough. Once he has established the first premise, the rest of the argument follows easily. The gist of the argument is that "the naturalist who sees that P(R/N & E) is low has a defeater for R, and for the proposition that his own cognitive faculties are reliable."[16]

A defeater is a belief that gives one a reason to give up a prior belief. For instance, let's suppose I'm walking on a sidewalk in downtown Seattle. I look across the street and see a man dashing into a bank who looks like President Barack Obama. In fact, he looks just like President Obama. I have read in the morning paper that President Obama is visiting Seattle on that day, and I don't know that Presidents don't just walk the streets of big cities unescorted. So, given what I know and have experienced, I might come to believe that I've just seen President Obama. But fifteen minutes later, when I tell my co-workers that I've seen the President, they explain that the President left Seattle hours before I thought I had seen him. They also explain that it's highly unlikely that he would be popping into a Bank of America by himself. Just then, someone turns on the television, and I see a live broadcast from Omaha, where the President is giving a speech. I now have several defeaters for my earlier belief that I had seen the President.

To cut to the chase, the entire evolutionary argument against naturalism goes like this:

1. P(R/N&E) is low.
2. Anyone who accepts (believes) N&E and sees that P(R/N&E) is low has a defeater for R.
3. Anyone who has a defeater for R has a defeater for any other belief she thinks she has, including N&E itself.
4. If one who accepts N&E thereby acquires a defeater for N&E, N&E is self-defeating and can't be rationally accepted.
5. *Conclusion: N&E can't rationally be accepted.*[17]

Remember that (1) is just an abbreviation for this sentence: *The probability that our cognitive faculties are reliable, given naturalism and Darwinian evolution, is low.* Or, a little less precisely: *If naturalism and Darwinian evolution are true, then our cognitive faculties probably aren't reliable.*

Here's how we might put the whole argument much less precisely: If you think clearly and carefully about the implications of Darwinian evolution and naturalism, you will see that, if they were true, then you should be skeptical about the reliability of your cognitive faculties, and so skeptical about your reasons for believing Darwinian evolution and naturalism are jointly true.

Although Plantinga doesn't mention it, many Darwinian naturalists argue that the entire human race may be hard-wired to believe grand falsehoods when it comes to religion, simply because those beliefs encourage selectively advantageous behavior, such as planning for the future, having children, altruism, and the like. So they not only admit that false beliefs can be paired with beneficial behaviors; they argue that this is exactly what has happened when it comes to some of our most important beliefs. But if natural selection has selected false beliefs even when those beliefs bear on our survival, what confidence could we really have in faculties that produce theoretical beliefs about what happened millions or billions of years ago, beliefs which have no obvious selective advantage?

As with Lewis's argument, this is not so much an argument against the truth of naturalism, as it is an argument against rationally believing that naturalism is true.

The only objection to Plantinga's argument that seems to me to have any weight is the anti-skeptical objection. Basically, such an objection simply rejects so-called "skeptical threat" arguments outright. Philosophers have been coming up with skeptical arguments for millennia to prove that change is impossible, that movement is impossible, that knowledge of the external world is impossible, that we might really be brains in vats, and so forth. But surely our constant interaction with the world, our agreement with others on many of the features of the world, should trump such skeptical trifles. Surely I'm much better off trusting that my cognitive faculties are reliable than trusting such abstract arguments which, if followed, would cause me to doubt what's right in front of my eyes.

I'm sympathetic to this objection. It seems unwise to abandon your common sense, and common experience, simply because of some skeptical argument from a philosopher. In order to follow the skeptical argument, after all, I'd have to rely on my cognitive faculties, and assume they're reliable. I'm sympathetic to the naturalist who replies: "C'mon, are we really supposed to become universal skeptics now?"

But I think this misses the point. Plantinga needn't argue that naturalists ought to doubt all their reasoning. His argument is that if naturalism were true, it's quite unlikely that our noetic and reasoning faculties would be reliably oriented toward the production of true beliefs. So believing in naturalism provides a defeater for believing it. Plantinga could just as well take the broad deliverances of our cognitive faculties as a given. He could stipulate that we should trust our sense that we have beliefs and intentions that have causal power. He could take for granted that our cognitive faculties are constructed to be generally reliable and aimed at the production of true beliefs, and that we can and should be persuaded of arguments in virtue of their propositional content and logical structure. But if naturalists want to join him, then they should now doubt the truth of naturalism.

The argument takes a probabilistic form of *modus tollens*: If A, then B. Not B, therefore, Not A.

- If Darwinian naturalism is true, then it's unlikely that we would have reliable cognitive faculties aimed at the production of true beliefs.
- It's not unlikely that our rational faculties are aimed at the production of true beliefs.
- *Conclusion: Therefore, Darwinian naturalism is probably not true.*

So the naturalist can take his choice: accept Darwinian naturalism and become a skeptic—about Darwinian naturalism and everything else—or renounce skepticism and doubt Darwinian naturalism.

Conclusion

Philosopher of science Michael Polanyi developed the idea of "tacit knowledge," and summarized it with the aphorism that "we know more than we can tell."[18] For instance, I know far more about my wife's appearance than I could distill into a description from memory. My description would be quite vague: "She's Caucasian, about 5 feet, 5 inches, and has brownish hair and blue eyes." But if you put her in a stadium with a thousand women, all of whom fit that description, I could easily pick her out. So clearly my knowledge of what she looks like goes far beyond what I can explicitly describe.

Some philosophical arguments seem to have the same quality. Lewis developed the argument from reason in a few pages in a popular book with no formal apparatus. And yet it has had staying power. I suspect that every year, thousands of people encounter the argument for the first time, and find that it moves them away from naturalism and toward theism. Lewis provided only a rough approximation of the argument, but it resides in a sweet spot. It is rigorous enough to be persuasive, and to tap into our rational and intuitive faculties, but not so technically precise as to be inaccessible. Victor Reppert was able to develop several distinct formal arguments from Lewis's original informal one. And Alvin Plantinga has constructed a formal version with unparalleled precision and rigor.

Each of these is an important task, but it is important that we not lose sight of Lewis's contribution in translating for millions of people one of the most compelling arguments against naturalism that has yet been devised.

Endnotes

1. C. S. Lewis Signature Classics (San Francisco: HarperOne, thirteenth edition, 2001).
2. "Version Vernacular," *The Christian Century* 75 (December 31, 1958): 515.
3. Victor Reppert, *C. S. Lewis's Dangerous Idea* (Downers Grove: InterVarsity Press, 2003), 87.
4. *Miracles* was first published in 1947. The revised edition came out in 1960. See C. S. Lewis, *Miracles: A Preliminary Study* (New York: Macmillan, 1960; first paperback edition, 1978).

5. Lewis, *Miracles* (1960), 14.
6. Ibid.
7. Ibid., 15.
8. Letter to William Graham, Down (July 3, 1881), in *The Life and Letters of Charles Darwin Including an Autobiographical Chapter*, edited by Francis Darwin (London: John Murry, Albermarle Street, 1887), vol. 1, 315–316.
9. The book was originally published in 1915. The most recent edition was published in 2000 (Seattle: Inkling Books) and edited by Michael W. Perry. Lewis included *Theism and Humanism* on a list of ten books that "did the most to shape" his "vocational attitude and… philosophy of life" published in *The Christian Century* on June 6, 1962.
10. Lewis used a version of the causal principle accessible to modern listeners. In his book on St. Thomas, G. K. Chesterton converted it into a characteristically witty aphorism: "It is absurd for the evolutionist to complain that it is unthinkable for an admittedly unthinkable God to make everything out of nothing, and then pretend that it is more thinkable that nothing should turn itself into everything." *St. Thomas Aquinas: The Dumb Ox* (New York: Random House, 1956), 145.
11. Alvin Plantinga, *Where the Conflict Really Lies: Science, Religion, & Naturalism* (New York: Oxford University Press), 310, footnote 4. He cites C. S. Lewis and philosopher Richard Taylor as offering precursors to his argument.
12. Ibid., 311.
13. Patricia Churchland, *Journal of Philosophy* 84 (October 1987): 548, emphasis in original. Quoted in Plantinga, 315.
14. Plantinga, *Where the Conflict Really Lies*, 321.
15. Ibid., 324.
16. Ibid., 339.
17. Ibid., 344–345.
18. He said this in many places, but perhaps the most detailed discussion is in *The Tacit Dimension* (Chicago: University of Chicago Press, reissue edition, 2009), 4. The book was originally published in 1964.

9

C. S. Lewis's Dangerous Idea Revisited

Victor Reppert

In this chapter, I will be considering what I have called elsewhere, "C. S. Lewis's Dangerous Idea"[1]—otherwise known as the "argument from reason." This argument, as we shall see, takes a number of forms, but in all instances it attempts to show that the necessary conditions of logical and mathematical reasoning, which undergird the natural sciences as a human activity, require the rejection of all broadly materialist worldviews.

I will begin by examining the nature of the argument, identifying the central characteristics of a broadly materialist worldview, and analyzing the prospects for a genuinely naturalistic alternative to a broadly materialist worldview. In so doing, I will examine the general problem of materialism, and how the argument from reason points to a single aspect of a broader problem. Second, I will examine the argument's history, including the famous dispute over it between C. S. Lewis and noted Roman Catholic philosopher Elizabeth Anscombe. In so doing, I will indicate how the argument from reason can surmount Anscombe's objections to it. I will also explain the transcendental structure of the argument. Finally, I will examine some popular objections, and show that these objections do not refute the argument.

The Nature of the Argument

Materialistic vs. Mentalist Worldviews

"In the beginning was the word." Although this statement, in its context, is laden with Christological implications, we can also use this statement to illustrate a central feature of certain worldviews, including Christian theism. The central idea is that fundamental to reality is that which is intelligible and rational. The metaphysical systems of Plato, Aristotle, and the Stoics, and of Hindu pantheism and Confucian philosophy as well, share this essential conception, as do the metaphysics of Spinoza and absolute idealism. The intelligible is fundamental to reality, the unintelligible or non-rational is, perhaps, a by-product of the created order, or an illusion caused by our own ignorance and lack of understanding. We might describe these worldviews as mentalistic worldviews. The mental is fundamental to reality; the non-mental is perhaps a creation, or perhaps a product of ignorance. Reality in mentalistic worldviews has a top-down character to it. The higher, mental levels create the lower levels, or the lower levels emanate from the higher levels. Alternatively, perhaps the lower levels are an illusion generated by the higher levels.

As science has progressed for the last few centuries, a move away from this kind of mentalistic worldview has emerged. According to broadly materialistic worldviews, it would be appropriate to say that in the beginning the word was not. Reason and intelligence are the by-product of centuries of evolution. As the higher primates evolved, they developed large brains, which provided them with true knowledge of the world around us, and this was an effective survival tool for them.

A good deal of debate within Western philosophy between worldviews has taken place between broadly mentalistic and broadly materialistic worldviews. (Sometimes people use the word naturalistic here, but for purposes of this discussion "broadly materialistic" will encompass all doctrines, that one could plausibly call naturalistic.) Christian theism has been the most popular, though by no means the only mentalistic worldview. Amongst broadly materialistic worldviews, there are options as well. Some proponents of materialistic views are eliminativist with respect to certain features of the mental lives that we common-sensically suppose ourselves to have. Other people in the materialistic camp maintain that we can account for many aspects of our mental lives through a reductive analysis of mind to the material. Still others believe that we can maintain a materialistic worldview by claiming that although we cannot reduce the mind to the material, the mind supervenes on the physical level.

In C. S. Lewis's critique of naturalism in the third chapter of *Miracles: A Preliminary Study*, Lewis presents what he calls "strict materialism," which he thinks can be refuted by a one-sentence quote from J. B. S. Haldane, on the one hand, and naturalism, which cannot be refuted so easily and requires a chapter-long treatment.[2] It seems to me, however, that the same people who in Lewis's time would have said, "We are naturalists but not materialists," would today describe themselves as materialists, given the considerably broader definition of materialism that is accepted today.

Lewis seems to imply that a real naturalism must be a form of determinism, claiming that a non-deterministic view implies a "subnatural" level which is a threat to naturalism.[3] Unlike Lewis, I am happy to allow that the advocate of a broadly materialist view can accept indeterminism at the subatomic level. However, this indeterminism is not a doorway for mentalistic explanations at the most basic level of analysis. Rather, this would have to be brute chance and nothing more.

I am convinced that a broadly materialist view of the world must possess three essential features:[4]

First, there must be a mechanistic base. Now by mechanistic I do not mean necessarily deterministic. There can be brute chance at the basic level of reality in a mechanistic world-view. However, the level of what I will call "basic physics" is free of purpose, free of meaning or intentionality, free of normativity, and free of subjectivity. If one is operating within a materialistic framework, then one cannot attribute purpose to what happens on the basic level. Purpose-talk may be appropriate for macro-systems, but it is a purpose that is ultimately the product of a purposeless basic physics. What something means cannot be an element of reality, as it appears at the most basic level. Moreover, there is nothing normative about basic physics. We can never say that some particle of matter is doing what it is doing because it ought to be doing that.

Rocks in an avalanche do not go where they go because it would be a good idea to go there. Finally, basic physics is lacking in subjectivity. The basic elements of the universe have no "points of view," and no subjective experience. Consciousness, if it exists, must be a "macro" feature of basic elements massed together.

Second, the level of basic physics must be causally closed. That is, if a physical event has a cause at time t, then it has a physical cause at time t. Even if that cause is not a determining cause; there cannot be something non-physical that plays a role in producing a physical event. If you knew everything about the physical level (the laws and the facts) before an event occurred, you could add nothing to your ability to predict where the particles would be in the future by knowing anything outside of basic physics.

Third, whatever is not physical, at least if it is in space and time, must supervene on the physical. Given the physical, everything else is a necessary consequence. In short, what the world is at bottom is a mindless system of events at the level of fundamental particles, and everything else that exists must exist in virtue of what is going on at that basic level. This understanding of a broadly materialist worldview is not a tendentiously defined form of reductionism; it is what most people who would regard themselves as being in the broadly materialist camp would agree with, a sort of "minimal materialism." Not only that, but I maintain that any world-view that could reasonably be called "naturalistic" is going to have these features, and the difficulties that I will be advancing against a "broadly materialist" worldview thus defined will be a difficulty that will exist for any kind of naturalism that I can think of.

The Argument from Reason and Natural Theology

We might ask the following question: In what sense is the argument from reason a piece of natural theology? The job of natural theology is supposed to be to provide epistemic support for theism. However, the argument from reason, at best, argues that the ultimate causes of the universe are mental and not physical. This is, of course, consistent with various worldviews other than traditional theism, such as pantheism or idealism.

It is a good idea to look at what happened in the case of the argument from reason's best-known defender, C. S. Lewis, to see how the argument contributed to his coming to belief in God. Lewis had been what philosophers of the time called a "realist," accepting the world of sense experience and science as rock-bottom reality. Largely through conversations with Owen Barfield, he became convinced that this worldview was inconsistent with the claims we make on behalf of our own reasoning processes. In response to this, however, Lewis became not a theist but an absolute idealist. It was only later that Lewis rejected absolute idealism in favor of theism, and only after that that he became a Christian.[5]

So, did the argument from reason that Lewis accepted make theism more likely in his mind? It certainly did. To understand why, consider the following argument:

1. Either at least some of the fundamental causes of the universes are more like a mind than anything else, or they are not.
2. If they are not, then it is either impossible or extremely improbable that reason should emerge.
3. All things being equal, worldviews that render it impossible or extremely improbable that reason should emerge should be rejected in favor of worldviews according to which it is not impossible and not improbable that reason should emerge.
4. Therefore, we have a good reason to reject all worldviews that reject the claim that the fundamental causes of the universe are more like a mind than anything else.

Strictly speaking, it is true that this argument does not establish theism. But it unquestionably makes theism more plausible by defeating one of its chief competitors (naturalism). To see why, suppose someone originally thinks that the likelihoods for various explanations of reality are as follows:

- Naturalism—50% likely to be true.

- Idealism—25% likely to be true.
- Theism—25% likely to be true.

Now further suppose that someone accepts a version of the argument from reason, and as a result, their belief in the likelihood of naturalism drops 30 percentage points. Those points then have to be divided amongst theism and idealism. Therefore, the epistemic status of theism is clearly enhanced by the argument from reason, if the argument is successful in defeating naturalism.

Some Problems with Materialism

The argument from reason is best understood as a response to what I call the general problem with materialism. The prime difficulty here is that the materialist holds, at the rock-bottom level, that the universe is an empty universe. As Lewis observes:

> At the outset, the universe appears packed with will, intelligence, life, and positive qualities; every tree is a nymph and every planet a god. Man himself is akin to the gods. The advance [of a materialist worldview] gradually empties this rich and genial universe, first of its gods, then of its colours, smells, sounds and tastes, finally of solidity itself as solidity was originally imagined. As these items are taken from the world, they are transferred to the subjective side of the account: classified as our sensations, thoughts, images or emotions. The Subject becomes gorged, inflated, at the expense of the Object. But the matter does not rest there. The same method which has emptied the world now proceeds to empty ourselves. The masters of the method soon announce that we were just mistaken (and mistaken in much the same way) when we attributed "souls" or "selves" or "minds" to human organisms, as when we attributed Dryads to the trees. Animism, apparently, begins at home. We, who have personified all other things, turn out to be ourselves mere personifications.[6]

When Lewis says the universe is empty from the standpoint of materialism, he means that it is empty of many of the things that are part of our normal existence. As I indicated, at the rock-bottom level, reality is free of normativity, free of subjectivity, free of meaning, and free of purpose according to the materialist worldview. All of these features of what makes life interesting for us are supposed to be late products of the struggle for survival. Yet it is these very features that cause many to criticize materialism as unable to account for all of the dimensions of human experience.

On the materialist view, for example, purpose itself does not seem justifiable. According to modern materialists, purpose must reduce to some Darwinian function. The purposeless motion of matter through space produced beings whose faculties perform functions that enhance their capacity to survive and pass on their genes. The physical is, on even the broadest of materialist views, a closed, non-purposive system, and any purpose that arises in such a world must be a byproduct of what, in the final analysis, lacks purpose. As Daniel Dennett puts it: "Psychology... must not explain intelligence in terms of intelligence, for instance assuming responsibility for the existence of intelligence to the munificence of an intelligent creator, or by putting clever *homunculi* at the control panels of the nervous system. If this were the best psychology could do, then psychology could not to the job assigned to it."[7]

In the final analysis, "purpose" exists in the world according to the materialist worldview not because there is, ultimately, any intended purpose for anything, but rather because things serve Darwinian functions. The claim that this type of analysis fails to adequately capture the kinds of purposiveness that exist provides the basis for arguments from design based on, for example, irreducible complexity.

Just as clearly, according to materialist worldviews, reality is free of subjectivity. The facts about the physical world are supposed to be objective facts that are not relative to anyone's subjective views. Once again, some have questioned how materialism can adequately account for our subjective inner states. Hence, we have arguments that point out that when all the physical facts about pain are given, we do not seem to have the grounding for, say, the state of what it is like to be in pain. We can imagine a possible world in which all the physical states obtain but whatever it is like to be in pain is missing. Arguments from consciousness arise from these considerations.

Similarly, normativity is absent at the physical level, giving rise to the notorious difficulty of getting an "ought" from an "is" in a materialist worldview. Let us begin with all the material facts about, let us say, the homicides of Ted Bundy. We can include the physical transformations that took place at that time, the chemical changes, the biology of the death process in each of these murders, the psychological state of the killer and his victims, the sociology of how membership in this or that social group might make one more likely to be a serial killer or a serial killer victim, etc. But from all of this, can we conclude that these homicides were morally reprehensible acts? We might know that most people believe them to be morally reprehensible acts, but whether they *are* reprehensible acts does not follow from any of this information. Materialism does not appear to offer a cogent answer.

Just as there are norms of morality, there are norms of rationality. Some patterns of reasoning are correct and others are not correct. If an argument is valid, we ought to accept its conclusion. If an argument is invalid, we should not. Some people have raised the question of how these norms of rationality can exist if materialism is true.

In addition, there is certainly no propositional content at the physical or material level. Yet it does seem to be possible to entertain a proposition. Here I am not even talking about belief (I think that p is true) or desire (I want p to be true) but just the process of entertaining the proposition and knowing what it means. It seems possible for propositions to be true or false, and for certain propositions to follow from others. But materialism does not seem to be able to account for this part of human experience.

Responses from the Defenders of Materialism

At this point, I am not endorsing the above objections to materialism. I am only saying that arguments of this sort are possible. One way for the skeptic to respond to these arguments is with an *error theory*:

We think there are objective moral norms, but we are mistaken because moral norms are subjective. Or we think conscious, subjective states really exist, but strictly speaking, they do not. As Susan Blackmore puts it: "each illusory self is a construct of the memetic world in which it successfully competes. Each selfplex gives rise to ordinary human consciousness based on the false idea that there is someone inside who is in charge."[8]

By referring to the self as "illusory," Blackmore is saying that what we ordinarily think of as consciousness does not exist. As we think of consciousness, we think of some center in which all mental states inhere. According to the *Stanford Encyclopedia of Philosophy*, consciousness has these characteristics: a first-person character, a qualitative character, a phenomenal structure, subjectivity, a self-perspectival organization, unity, intentionality, and dynamic flow.[9] Error theories of consciousness, such as Blackmore's, instead of showing how these aspects of consciousness can exist in a materialist world, suggest we are mistaken in thinking that these elements really exist.

Besides error theories, there are two other types of responses defenders of materialism generally use to criticize the family of arguments I have presented above. They use *reconciliation* objections if they suppose that the item in question can be fitted within a materialist ontology. And they use *inadequacy* objections to argue that whatever difficulties there may be in explaining the matter in materialist terms, it does not get us any better explanations if we accept some mentalistic worldview like theism.

We can see this typology at work in responses to the argument from objective moral values. Materialist critics of the moral argument can argue that there is really no objective morality; or they can say objective morality is compatible with materialism; or they can use arguments like the Euthyphro dilemma to argue that whatever we can't explain about morality in materialist terms cannot better be explained by appealing to nonmaterial entities such as God.

However, it is important to notice something about the defenders of materialism. They not only believe that the world is material. They also perforce believe that the truth about that material world can be discovered, and is being discovered, by people in the sciences. Furthermore, they believe there are philosophical arguments that ought to persuade people to eschew mentalistic worldviews in favor of materialistic ones. And they think we can better discover the nature of the world by observation and experimentation than by reading tea leaves.

But here arguments from reason have a "home court" advantage. By their very nature, arguments from reason appeal to the necessary conditions of rational thought and inquiry. Thus, they have a built-in defense against error-theory responses. If there is no truth, materialists have no basis to claim that materialism itself is true. If there is no mental causation, then materialists cannot claim that our beliefs ought to be based on supporting evidence. If there are no logical laws, then materialists cannot say that the argument from evil is a good argument. If our rational faculties as a whole are unreliable, then materialists cannot argue that religious beliefs are formed by irrational belief-producing mechanisms. As a result, arguments from reason have what I call a transcendental impact—that is, appeal to things that, if denied, undermine the most fundamental convictions of philosophical materialists. There cannot be a scientific proof that scientists do not exist; that would undermine the scientific enterprise that constitutes the very foundation of materialism.

The Reality of Rational Inference

THE ARGUMENT FROM REASON FOCUSES on cases where we infer one proposition from another proposition. Many have argued, however, that we can know certain things directly without using inferences. For example, I can have a justified belief that my eyeglasses are here on my computer table without drawing any inferences at all, but rather, just by perceiving my glasses. I should add that this "direct realist" view of perception is by no means universal amongst philosophers; there are many who maintain that what we are directly aware of are "sense data" and that we infer physical objects from sense data. But assuming the truth of the "direct realist" view, what impact does it have on the argument from reason?

John Beversluis, for one, seems to think that the "direct realist" view defeats the argument from reason as articulated by C. S. Lewis. Beversluis contends that Lewis's argument from reason relies on an inference-from-sense-data theory of the knowledge of physical objects, and fails unless that theory is defensible.[10] However, even if Lewis himself held such a position (and I think he probably did, though the evidence is less than crystal-clear), Lewis's argument does not *need* to rely on inferential theories of sensory knowledge.

Here is why: Even if we concede that there is some knowledge that is not inferred from sensations and therefore does not depend on the validity of reasoning, advocates of materialistic worldviews are committed to the existence of *at least some* inferential knowledge. They must hold that *scientists* make rational and mathematical inferences; and they must hold that they themselves accept materialism because there is good reason to believe it. Even if we accept the "direct realist" view with respect to physical objects, there is a whole lot of rationally inferred knowledge that no materialist can dare deny, on pain of undermining both science and naturalism.

The claim that rational inferences are essential to the possibility of science even though there may be other sources of justified beliefs is sometimes overlooked by persons who respond to the argument from reason, so I will name this argument the "critical subset argument." Thus I do not need to agree with C. S. Lewis that "[a]ll possible knowledge… depends on the validity of reasoning."[11] (italics mine). Even if only a subclass of knowledge depends on the validity of reasoning, the subclass of knowledge that does is a subclass critical to the materialist's enterprise.

Materialists maintain, of course, that what is real are the sorts of things that lend themselves to scientific analysis. But they cannot escape believing that there are scientists and mathematicians whose minds are capable of performing those scientific analyses. Consider, for example, a doctrine I call "hyper-Freudianism," the view that all beliefs are the product of unconscious drives, and that no one believes anything they believe for the reasons that they think they believe it. An atheist could say of the theist, "You think you believe in God because of the arguments of Christian apologists, but you really believe it because you are searching for a cosmic father figure to calm your fears." Alternatively, a theist can say, "You think you are an atheist because of the evidence of evolution and the problem of evil, but I know that you just want to kill your father." Of course, this kind of objection can be pushed still further to include all beliefs. However, that is just the trouble. If the objection is pushed that far, then it has to be extended to the belief in hyper-Freudianism itself. If someone tries to present evidence for hyper-Freudianism, he is doing something that can only be done if hyper-Freudianism is false.

Similarly, consider the role of mathematics in science. Mathematical inferences were critical in making it possible for Newton to discover the theory of gravity and Einstein to discover relativity. If we believe that natural science gets the truth about the world, then we must not deny that mathematical inferences exist. More generally, if we are persuaded by scientific arguments about any topic at all, then we must not accept a position that entails that no one is ever persuaded by an argument.

The History of the Argument

Antecedents of Lewis's Argument from Reason

THE ARGUMENT FROM REASON DID not originate with C. S. Lewis. Something like it can be traced all the way back to Plato, and Augustine had an argument that said that our knowledge of eternal and necessary truths showed that God exists. Descartes maintained that the higher rational processes of human beings could not be accounted for in materialistic terms, and while Kant denied that these considerations provided adequate proof of the immortality of the soul, he did think they were sufficient to rule out any materialist account of the mind.[12] However, naturalism or materialism as a force in Western culture increased considerably in 1859 when Charles Darwin published the *Origin of Species*.

The earliest post-Darwinian presentation of the argument from reason that I am familiar with, and one that bears a lot of similarities to Lewis's argument, is found in Prime Minister Arthur Balfour's *The Foundations of Belief* (1895).[13] Although Lewis never mentions *The Foundations of Belief* in his writings, his father Albert owned a copy, and Lewis eventually incorporated the copy into his own personal library.[14] Perhaps more significantly, Lewis credited Balfour's subsequent book *Theism and Humanism* (1915)[15] as one of the ten books that influenced him the most, commenting in another place that it is "a book too little read."[16] In that second book, Balfour basically repeated his argument from reason. Another thinker from the early twentieth century who popularized a version of the argument was J. B. Pratt in his book *Matter and Spirit* (1922), which presented the argument from reason as an argument for mind-body dualism.[17]

Lewis's First Edition Argument

IN THE first edition of *Miracles*, Lewis presents the version of the argument from reason that philosopher Elizabeth Anscombe criticized. We can formalize it as follows:

- No thought is valid if it can be fully explained as the result of irrational causes.
- If materialism is true, all beliefs can be explained in terms of irrational causes.
- Therefore, if materialism is true, no thought is valid.
- If no thought is valid, the thought, "materialism is true," is not valid.
- Therefore, if materialism is true, the belief "materialism is true" is not valid.
- A thesis whose truth entails the invalidity of the belief that it is true ought to be rejected, and its denial ought to be accepted.
- Therefore, materialism ought to be rejected, and its denial accepted.[18]

Anscombe's critique is significant because of the way in which it forced Lewis to develop and improve his arguments. We will examine three challenges Anscombe put to Lewis's argument to see how the argument needs to be refined to meet the challenges.

Anscombe's First Objection: Irrational vs. Nonrational

ANSCOMBE QUESTIONED whether it was correct for Lewis to talk about physically caused events as having *irrational* causes. Irrational beliefs, one would think, are beliefs that are formed in ways that conflict with reason: wishful thinking, for example, or fallacious arguments. On the other hand, when we speak of a thought having a non-rational cause, we need not be thinking that there is any conflict with reason.[19]

However, this distinction—while legitimate—is hardly sufficient to refute Lewis's argument. Remember, a materialist philosopher not only believes that some beliefs are justified, a materialist, if she thinks that science is true, thinks that some people do draw correct logical and mathematical inferences. While not all justified beliefs are inferred from other beliefs, a contemporary materialist is not in a position to maintain that beliefs are formed as a result of rational inferences.

For that reason, it is possible to restate Lewis's argument in such a way that it does not make reference to irrational causes, and indeed in Lewis's revised chapter the phrase "irrational causes" does not appear. Here is a restated version:

- No belief is rationally inferred if it can be fully explained in terms of nonrational causes.
- If materialism is true, all beliefs can be fully explained in terms of nonrational causes.
- Therefore, if materialism is true, no belief is rationally inferred.
- If any thesis entails the conclusion that no belief is rationally inferred, it should be rejected and its denial accepted.
- Therefore, materialism should be rejected and its denial accepted.

I could go a bit farther in defending Lewis against Anscombe's critique here. In my 1989 essay, "The Lewis-Anscombe Controversy: A Discussion of the Issues," and in *C. S. Lewis's Dangerous Idea*, I discussed Anscombe's insistence that Lewis distinguish between irrational causes and non-rational causes.[20] Irrational causes would be things like being bitten by a black dog as a child causing you to believe that all black dogs are dangerous. Nonrational causes are physical events or physical causes. Interestingly enough, in a passage from *The Abolition of Man* (written several years before Anscombe's challenge), Lewis seems to make the distinction Anscombe was looking for: "Now the emotion, thus considered by itself, cannot be either in agreement or disagreement with Reason. It is irrational not as a paralogism is irrational, but as a physical event is irrational: it does not rise even to the dignity of error."[21]

Now, in this passage Lewis draws the exact distinction on which Anscombe insisted. The only difference here is that Lewis distinguishes two senses of the term "irrational" instead of distinguishing between irrational and nonrational. Was Lewis's usage of the term "irrational" in this context wrong? Going to a dictionary definition of "irrational," I think not. The very first entry for "irrational" at www.dictionary.com is "without the faculty of reason; deprived of reason."[22] Nevertheless, in the revised edition of *Miracles* Lewis changed his term from "irrational" to "nonrational" to accommodate Anscombe's criticism. However, physical causes are by nature irrational causes in the sense presented in the dictionary definition, so his use of "irrational" was not mistaken.

Anscombe's Second Objection:
Paradigm Cases and Skeptical Threats

ANSCOMBE ALSO objected to Lewis's claim that if naturalism were true, reasoning would not be valid. She asks, "What can you mean by 'valid' beyond what would be indicated by the explanation you would give for distinguishing between valid and invalid, and what in the naturalistic hypothesis prevents that explanation from being given and from meaning what it does?"[23] This is a paradigm case argument, and the point is this: We can ask whether a particular argument is a good one, but does it really make sense to argue that reasoning might itself be invalid? Anscombe maintains that since the argument that some particular piece of reasoning is invalid involves contrasting it with some other kinds of reasoning that are valid, the question "Could reasoning really be valid?" is really a nonsense question.

One way of using the argument from reason would be to use it as a skeptical threat argument. The idea is that if naturalism is true we will be unable to refute skeptical arguments against reasoning in general. The problem here is that it is far from clear that anyone, naturalist or not, can refute skepticism about reasoning, nor is it considered any great merit for any metaphysical theory that it would be possible to refute this kind of thoroughgoing skepticism. And, if we need to refute skepticism in order to accept some worldview, it is not at all clear that theism will do that either. If we use our theistic beliefs to defend the basic principles of reasoning, then we would have to formulate that into an argument and then *presuppose* our ordinary canons of logical evaluation in the presentation of that very argument, thereby begging the question.

In my previous writings on the subject, I present the argument from reason instead as a best explanation argument. One should assume, at least to begin with, that human beings do reach true conclusions by reasoning, and then try to show, given the fact that people do reach true conclusions by reasoning, that this is best explained in terms of a theistic metaphysics as opposed to a naturalistic metaphysics. Now if we present the argument in this way and an opponent comes along and says, "I see that your argument presupposes that we have beliefs. I do not think we do, so your argument fails," we can reply to him thus: "If there are no beliefs, then you do not believe what you are saying. Consequently, the status of your own remarks as *assertions* is called into question by your own thesis that there are no beliefs, and this is going to end up having a devastating effect on the very sciences on which you base your arguments." Presenting the argument in this way, it seems to me, gets around the problems based on the paradigm case argument.

William Hasker, however, while previously endorsing the gist of my claim that the argument should be a best explanation argument rather than a skeptical threat argument, offered another suggestion in his mostly friendly response to me in *Philosophia Christi*. He wrote:

> However, if the skeptical threat strategy claims too much for the Argument from Reason, there is a danger that the Best Explanation strategy may claim too little. On the face of it, this strategy seems to invite the following kind of response: "It may be true that we naturalists have not, so far, produced a satisfying explanation for the process of rational inference. But there is nothing especially surprising or alarming about this fact. Finding good scientific explanations is hard work and often takes considerable time, and the relevant sciences are still in their infancy. We must simply be prepared to wait a bit longer, until we reach the stage where the desired explanations can be developed."[24]

Hasker then made the following recommendation:

> The objection is not merely that naturalism has not yet produced an explanation of rational inference and the like, as though this were a deficiency that could be remedied by another decade or so of scientific research. The problem is that the naturalist is committed to certain assumptions that preclude in principle any explanation of the sort required. The key assumptions are three in number: mechanism (the view that fundamental physical explanations are nonteleological), the causal closure of the physical domain, and the supervenience of the mental on the physical. So long as these assumptions remain, no amount of ingenious computer modeling can possibly fill the explanatory gap. In order to bring out this feature of the situation, I propose that the first two stages of the Argument from Reason are best viewed as a transcendental argument in roughly the Kantian sense: They specify the conditions which are required for experience of a certain sort to be possible—in this case the kind of experience found in the performance of rational inference.[25]

I have already discussed the transcendental impact of the arguments from reason, and I think Hasker's suggestion is a good one. So what are the conditions required for rational inference to take place? Consider the following list:

- States of mind have a relation to the world we call intentionality, or about-ness.
- Thoughts and beliefs can be either true or false.
- Humans can be in the condition of accepting, rejecting, or suspending belief about propositions.
- Logical laws exist.
- Human beings are capable of apprehending logical laws.
- The state of accepting the truth of a proposition plays a crucial causal role in the production of other beliefs, and the propositional content of mental states is relevant to the playing of this causal role.
- The apprehension of logical laws plays a causal role in the acceptance of the conclusion of the argument as true.
- The same individual entertains thoughts of the premises and then draws the conclusion.
- Our processes of reasoning provide us with a systematically reliable way of understanding the world around us.[26]

Unless all of these conditions are true, it is incoherent to argue that one should accept naturalism based on evidence of any kind. Nor would it be possible to accept the claim that one should accept evolution as opposed to creationism because there is so much evidence for evolution. Nor could one argue that one should be supremely confident that use of the scientific method will result in an accurate understanding of reality. Unless *all* these conditions are true, there are no scientists and no one is using the scientific method. These conditions provide a transcendental justification for rational inference. That is, once one grants that there is at least one person who has made a rational inference (such as a mathematical calculation), one also grants that these specified conditions must be true.

Anscombe's Main Objection: The Ambiguity of "Why," "Because," and "Explanation"

THE THIRD and main Anscombe objection to Lewis's argument is that Lewis fails to distinguish between different senses of the terms "why," "because," and "explanation." There are, Anscombe suggests, four explanation types that have to be distinguished:

1. Naturalistic causal explanations, typically subsuming the event in question under some physical law.
2. Logical explanations, showing the logical relationship between the premises and the conclusion.
3. Psychological explanations, explaining why a person believes as he/she does.
4. Personal history explanations, explaining how, as a matter of someone's personal history, that person came to hold a belief.

Anscombe suggests that arguments of different types can be compatible with one another. Thus, a naturalistic causal explanation might be a complete answer to one type of question with respect to how someone's belief came to be what it was, but that explanation might be compatible with a "full" explanation of a different type.[27]

Now what is interesting is that Lewis, in reformulating his own argument, not only draws the distinctions on which Anscombe had insisted; he actually makes these distinctions the centerpiece of his revised argument. He makes a distinction between cause-and-effect relations on the one hand, and ground-and-consequent relations on the other. Cause-and-effect relations say how a thought was produced, but ground-and-consequent relations indicate how thoughts are related to one another logically. However, in order to allow for rational inference, there must be a combination of ground-consequent and cause-effect relationships, which, Lewis says, cannot exist if the world is as the naturalist says that it is.

Claiming that a thought has been rationally inferred is a claim about how that thought was caused. Any face-saving account of how we come to hold beliefs by rational inference must maintain that "[o]ne thought can cause another thought not by being, but by being seen to be, a ground for it."[28]

According to Lewis, there are a number of features of thoughts as they occur in rational inference that set them apart from other events that occur in the world:

> Acts of thinking are no doubt events, but they are special sorts of events. They are "about" other things and can be true or false. Events in general are not "about" anything and cannot be true or false... Hence acts of inference can, and must be considered in two different lights. On the one hand they are subjective events in somebody's psychological history. On the other hand, they are insights into, or knowings of, something other than themselves.[29]

So here, we already have three features of acts of thinking as they occur in rational inference. First, these thoughts have to be about something else. Second, they can be true or false. Third, their propositional contents must cause other thoughts to take place. But there is more:

> What from the first point of view is a psychological transition from thought A to thought B, at some particular moment in some particular mind is, from the thinker's point of view, a perception of an implication (if A, then B). When we are adopting the psychological point of view we may use the past tense, "B followed A in my thoughts." But when we assert the implication we always use the present—"B *follows from* A." If it ever "follows from" in the logical sense it does so always. Moreover, we cannot reject the second point of view as a subjective illusion without discrediting human knowledge.[30]

So now, in addition to the three features of thoughts as they occur in rational inference, we can add a fourth, that is, that the act of inference must be subsumed under a logical law. Moreover, the logical law according to which one thought follows another thought is true always. It is not local to any particular place or time; indeed laws of logic obtain in all possible worlds.

Lewis then argues that an act of knowing "is determined, in a sense, by what is known; we must know it to be thus because it is thus."[31] *P*'s being true somehow brings it about that we hold the belief that *P* is true. Ringing in my ears is a basis for knowing if a ringing object causes it; it is not knowledge if it is caused by tinnitus:

> Anything that professes to explain our reasoning fully without introducing an act of knowing thus solely determined by what it knows, is really a theory that there is no reasoning. But this, as it seems, is what Naturalism is bound to do. It offers what professes to be a full account of our mental behaviour, but this account, on inspection, leaves no room for the acts of knowing or insight on which the whole value of our thinking, as a means to truth, depends.[32]

If a broadly materialist worldview is true, how is it possible for our acts of rational inference to occur because reality has a feature that corresponds to that inferential process? That is the question Lewis thinks a materialist cannot answer.

Unlimited Explanatory Compatibility and the Noncausal View of Reason

THIS IS a point at which Anscombe, in her brief response to Lewis's revised argument, objects, claiming that Lewis did not examine the concept of "full explanation" that he was using. Anscombe had expounded a "question relative" conception of what a "full explanation" is; a full explanation gives a person everything they want to know about something. What this appears to result in is the idea of an unlimited explanatory compatibilism. It is further supported if one accepts, as Anscombe did when she wrote her original response to Lewis, the Wittgensteinian doctrine that reasons-explanations are not causal explanations at all. They are simply sincere responses that are elicited from a person when he is asked what his reasons are. As Anscombe puts it: "It appears to me that if a man has reasons, and they are good reasons, and they genuinely are his reasons, for thinking something—then his thought is rational, whatever causal statements can be made about him."[33]

Keith Parsons adopted essentially the same position in response to my version of the argument from reason when he wrote:

> My own (internalist) view is that if I can adduce reasons sufficient for the conclusion Q, then my belief that Q is rational. The causal history of the mental states of being aware of Q and the justifying grounds strike me as quite irrelevant. Whether those mental states are caused by other mental states, or caused by other physical states, or just pop into existence uncaused, the grounds still justify the claim.[34]

However, the claim that reasons-explanations are not causal explanations at all seems to me to be completely implausible. As Lewis puts it:

> Even if grounds do exist, what have they got to do with the actual occurrence of belief as a psychological event? If it is an event it must be caused. It must in fact be simply one link in a causal chain which stretched back to the beginning and forward to the end of time. How could such a trifle as lack of logical grounds prevent the belief's occurrence and how could the existence of grounds promote it?[35]

It is interesting that we often describe evidence in almost physical terms. Lewis says, for example, "I am not asking anyone to believe in Christianity if his best reasoning tells him the weight of the evidence is against it."[36] Our common-sense use of the term "weight" here, which treats evidence as if it had kinetic energy, is a strong testimony to common sense's commitment to the causal role of reasons.

If you were to meet a person, call him Steve, who could argue with great cogency for every position he held, you might be inclined to consider him a very rational person. However, suppose that on all disputed questions Steve rolled dice to fix his positions permanently and then used his reasoning abilities only to generate the best-available arguments for those beliefs selected in the above-mentioned random method. I think that such a discovery would prompt you to withdraw from him the honorific title "rational." Clearly, we cannot answer the question of whether or not a person is rational in a manner that leaves entirely out of account the question of how his or her beliefs are produced and sustained.

There do seem to be limits on explanatory compatibility. Consider how we explain how presents came to appear under the Christmas tree.

If we accept the explanation that, in spite of the tags on the presents that say Santa Claus, Mom and Dad in fact put the presents there, this would of course conflict with the explanation in terms of the activity of Santa Claus. An explanation of disease in terms of microorganisms is incompatible with an explanation in terms of a voodoo curse. In fact, naturalists are the first to say, "We have no need of that hypothesis," if a materialistically acceptable explanation can be given where a supernatural explanation had previously been accepted.

Further, explanations, causal or non-causal, involve *ontological commitments*. That which plays an explanatory role is supposed to *exist*. Therefore, if we explain the existence of the presents under the Christmas tree in terms of Santa Claus, I take it that means that Santa Claus exists in more than just a non-realist "Yes, Virginia" sense. Anscombe seemed to think that all that is involved in naturalism is that every event can be given a naturalistic explanation. But naturalism and materialism are typically defined in ontological terms. For example, the main page of the Internet Infidels website approvingly quotes philosopher Paul Draper's description of naturalism as

> the "hypothesis that the physical world is a closed system" in the sense that "nothing that is not a part of the natural world affects it." As such, "naturalism implies that there are no supernatural entities," such as gods, angels, demons, ghosts, or other spirits, "or at least none that actually exercises its power to affect the natural world."[37]

What this means is that even if reasons-explanations do not exclude physical explanations, even if reasons-explanations are somehow not causal explanations, the naturalist is not out of the woods. The materialist maintains that the universe, at its base, is governed by blind matter rather than reasons. So if reasons-explanations are true, we still need to know why they are true and why reasons exist in a world that is fundamentally non-rational.

In her final response to Lewis, Anscombe made two further complaints. The first was that Lewis did not repair the concept of a "full explanation," and that he did not adequately explore the idea of "an act of knowing solely determined by what is known." On the other hand, she admitted that "We haven't got an answer" to the question "Even if grounds do exist, what have they got to do with the actual occurrence of belief?"[38]

Because of the ambiguities connected with the idea of "full explanation," my own development of the argument from reason has avoided talk about full explanations, but has instead focused on the ideas of mechanism, causal closure, and supervenience. Given these three doctrines, it seems as if some kinds of explanations face the prospect of being ruled out. Even the most non-reductive forms of materialism maintain that there can be only one kind of causation in a physicalist world, and that is physical causation. It is not enough simply to point out that we can give different "full" explanations for the same event. Of course, we can. Nevertheless, given the causal closure thesis of materialism there cannot be causal explanations that require non-materialist ontological commitments. The question that is still open is whether the kinds of mental explanations required for rational inference are compatible with the limitations placed on causal explanations by materialism. If not, then we are forced to choose between saying that there are rational inferences and accepting materialism. However, materialism is invariably presented as the logical conclusion of a rational argument. Therefore, the choice will have to be to reject materialism.

Anscombe's second complaint was to insist that Lewis needed to clarify his conception of "an act of knowing" determined by "what is known." Now in one sense this can be made clear by reflecting on the correspondence theory of truth. If I am in the state of believing that the cat is on the mat, which corresponds to some state of the world, what makes it a case of knowledge instead of just a lucky guess or something like that? If the cat is on the mat, and I know that, then somehow the cat's being on the mat has to play some role in producing the belief in me that the cat is on the mat. Otherwise, the cat could be anywhere but on the mat and I would still think it was there. However, in place of this concept, I prefer to employ the set of transcendentally established conditions of reason and science that I listed above.

How Could Reason Emerge?

LEWIS MAINTAINS THAT IF WE acquired the capability for rational inference in a naturalistic world it would have to have arisen either through the process of evolution or because of experience. Regarding the first option, Lewis argues that evolution will always select for improved responses to the environment, and evolution could do this without actually providing us with inferential knowledge:

> natural selection could operate only by eliminating responses that were biologically hurtful and multiplying those which tended to survival. But it is not conceivable that any improvement of responses could ever turn them into acts of insight, or even remotely tend to do so. The relation between response and stimulus is utterly different from that between knowledge and the truth known... our psychological responses to our environment—our curiosities, aversions, delights, expectations—could be indefinitely improved (from the biological point of view) without becoming anything more than responses. Such perfection of the non-rational responses, far from amounting to their conversion into valid inferences, might be conceived as a different method of achieving survival—an alternative to reason.[39]

With regard to experience, Lewis points out that experience might cause us to expect one event to follow another, but to logically deduce that we *should* expect one effect to follow another is not something that could be given by experience. Experience can show us that A succeeds B. It cannot show us that A follows necessarily from B. As Lewis writes: "My belief that things which are equal to the same thing are equal to one another is not at all based on the fact that I have never caught them behaving otherwise. I see that it 'must' be so."[40]

Lewis then makes his case that theism, at least, can avoid the kinds of problems that the Naturalist faces:

> the Theist... is not committed to the view that reason is a comparatively recent development moulded by a process of selection which can select only the biologically useful. For him, reason—the reason of God—is

older than Nature, and from it the orderliness of Nature, which alone enables us to know her, is derived. For him, the human mind in the act of knowing is illuminated by the Divine reason. It is set free, in the measure required, from the huge nexus of non-rational causation; free from this to be determined by the truth known. And the preliminary processes within Nature which led up to this liberation, if there were any, were designed to do so.[41]

The Argument Since Lewis

LEWIS'S ARGUMENT HAS BEEN ECHOED in various places since the debate with Anscombe. One of the lesser-known developments in the argument was a defense of Lewis against Anscombe in the book *Christian Theology and Natural Science* by Eric Mascall.[42] The argument has appeared in various other books and essays as well. J. R. Lucas claims that the central argument of his book *Freedom of the Will* was inspired by Lewis, although Lucas enlists Godel's theorem to defend his version of the argument.[43] Perhaps the best defense of a Lewis-like argument from reason between the 1960s and the 1990s came in William Hasker's essay, "The Transcendental Refutation of Determinism."[44] John Beversluis, in a book highly critical of all of Lewis's arguments, was the first to use the phrase "argument from reason" to designate this particular line of reasoning.[45] My own efforts with respect to the argument were begun when I wrote my doctoral dissertation in defense of it, entitled "Physical Causes and Rational Belief: A Problem for Materialism?"[46] The second chapter of my dissertation became my original essay on the Lewis-Anscombe controversy. Shortly thereafter Alvin Plantinga published his first defense of what came to be known as the evolutionary argument against naturalism, which first appeared in his book *Warrant and Proper Function* and was developed in more detail in *Warranted Christian Belief*.[47] In 1999 William Hasker published a chapter-long defense of the argument entitled "Why the Physical Isn't Closed," in his book *The Emergent Self*.[48] I returned to the defense of the argument starting in 1999 with an exchange in *Philo* that spilled over into *Philosophia Christi*.[49] Then my book *C. S. Lewis's Dangerous Idea: In Defense of the Argument from Rea-*

son was published in 2003, and I have engaged in further development and discussion of the argument since.[50] In addition, other defenders of the argument have emerged, such as Darek Barefoot,[51] Angus Menuge,[52] and Michael Rea.[53]

The Elements of Reason for C. S. Lewis

One aspect of my own discussion of the argument that has, I think, influenced the discussion of the argument the most is my subdivision of the argument from reason into six subarguments. In examining the argument I found that the argument focused on different elements of the reasoning process, and that one could find difficulties for naturalism at more than one step along the way.

Perhaps Lewis himself also noticed that there are different elements to the process of rational inference. Consider this description of inference, which, interestingly enough, occurs in a critique of pacifism, not in a presentation of the argument from reason:

> Now any concrete train of reasoning involves three elements: *Firstly*, there is the reception of facts to reason about. These facts are received either from our own senses, or from the report of other minds; that is, either experience or authority supplies us with our material. But each man's experience is so limited that the second source is the more usual; of every hundred facts upon which to reason, ninety-nine depend on authority. *Secondly*, there is the direct, simple act of the mind perceiving self-evident truth, as when we see that if A and B both equal C, then they equal each other. This act I call intuition. *Thirdly*, there is an art or skill of arranging the facts so as to yield a series of such intuitions, which linked together produce, a proof of the truth of the propositions we are considering. This in a geometrical proof each step is seen by intuition, and to fail to see it is to be not a bad geometrician but an idiot. The skill comes in arranging the material into a series of intuitable "steps". Failure to do this does not mean idiocy, but only lack of ingenuity or invention. Failure to follow it need not mean idiocy, but either inattention or a defect of memory which forbids us to hold all the intuitions together."[54]

So Lewis isolates three steps in the reasoning process: 1) The reception of facts to think about; 2) The perception of a self-evident truth of rule that permits the inference; and 3) Arranging the fact to prove a conclusion. Sometimes, in developing the argument from reason, advocates point out the difficulty the naturalist has in giving an account of how it is a thought can be *about* something. This aspect of thought, which philosophers since Brentano have called *intentionality*, has often been thought to be profoundly problematic for the philosophical naturalist. The next step in the process seems problematic as well—how it is that purely natural creatures completely embedded in the space-time continuum could know not only that something is true, but that it must be true. Our physical senses might perceive what is, but how could physical beings know what aspects of what they experienced could not be otherwise? Finally, what happens when we arrange statements to prove a conclusion? It seems that our understanding of the propositional content of one statement has to be the deciding factor in our being able to draw the conclusion. As Lewis asked in his revised chapter, "Even if grounds do exist, what exactly have they got to do with the actual occurrence of belief as a psychological event?" Hence it looks as if the naturalist, in order to affirm the existence of rational inference, must accept the existence of mental causation in which the state of accepting the content of one statement causes the acceptance of the content of another statement. How mental causation can fit into a naturalistic world has been widely regarded as a problem.

Two Popular Objections

1. The Argument from Computers

SOME PEOPLE think it is easy to refute any argument from reason just by appealing to the existence of computers. Computers, according to the objection, employ reason. They are undeniably physical systems that are also rational. So whatever incompatibility there might be between mechanism and reason must be illusory.

However, in the case of computers, the compatibility is the result of mental states in the background that deliberately create this compatibility. Thus, the chess computer Deep Blue was able to defeat the world champion Garry Kasparov in their 1997 chess match. However, Deep Blue's ability to defeat Kasparov was not the exclusive result of physical causation, unless the people on the programming team (such as Grandmaster Joel Benjamin) are entirely physical results of physical causation. To assume that, however, is to beg the question against the advocate of the argument from reason. As Hasker points out: "Computers function as they do because they have been constructed by human beings endowed with rational insight. A computer, in other words, is merely an extension of the rationality of its designers and users; it is no more an independent source of rational insight than a television set is an independent source of news and entertainment."[55]

The argument from reason says that reason cannot emerge from a closed, mechanistic system. The computer is, narrowly speaking, a mechanistic system, and it does "follow" rational rules. But not only was the computer made by humans, the framework of meaning that makes the computer's actions intelligible is supplied by humans. As a set of physical events, the actions of a computer are just as subject as anything else to the indeterminacy of the physical. If a computer plays the move Rf6, and we see it on the screen, it is our perception and understanding that gives that move a definite meaning. In fact, the move has no meaning to the computer itself; it only means something to persons playing and watching the game. Suppose we lived in a world without chess, and two computers were to magically materialize in the middle of the Gobi desert and go through all the physical states that the computers went through the last time Fritz (a computer chess program) played Shredder (another computer chess program). If that were true, these two computers would not be playing a chess game at all, since there would be no humans around to impose the context that made those physical processes a chess game and not something else. Hence I think that we can safely regard the computer objection as a red herring.

2. God of the Gaps

ANOTHER ARGUMENT frequently advanced against virtually any piece of natural theology is the God of the Gaps charge. In fact, this is one of the most popular items in the atheist playbook. We know from the history of science that many things which were thought in the past to require an explanation in terms of divine agency are now known to have naturalistic explanations. Rainbows, for example, were once thought to have been put in the sky as a sign, but we now know that they can be naturalistically explained in terms of light refraction. Various biological systems show a harmony between means and ends which in the past was cannon fodder for the design argument, but is now explicable in terms of random variation and natural selection. So if there is something that we think cannot be explained in physical terms, just give scientists some time, and they'll figure it out sooner or later.

An instance where the God of the Gaps objection appears strong is in the case of Newton's account of the orbits of the planets. His theory would have expected the orbits to go somewhat differently from the way they go, and so he postulated God as the one who keeps the planets in line. Laplace later developed a theory that didn't require this kind of divine tinkering, and when asked about Newton's theistic theory he said "I have no need of that hypothesis."

However, I am not sure that every argument that points to an explanatory difficulty for the naturalist can be effectively answered with a "God of the Gaps" charge. Consider, for example, being at a dinner party with someone who is given a large amount of water and creates from it an equal volume of wine. (It tastes like really good wine, not that California cheap stuff.) Can we reasonably say that in this situation we just have a gap in our understanding of natural processes? As Robert Larmer points out, our understanding of how wine is made is precisely what makes it so difficult to explain naturalistically:

> Claims regarding events traditionally described as miracles and claims regarding the origin and development of life are where "God of the gaps" arguments are most commonly met. In the case of events traditionally

described as miracles, it seems very evident that our increased knowledge of how natural causes operate has not made it easier, but more difficult, to explain such events naturalistically. The science underlying wine-making is considerably more advanced today than it was in first century Palestine, but our advances have made it even more difficult to explain in terms of natural causes how Jesus, without any technological aids, could, in a matter of minutes, turn water into high quality wine. Indeed, it is the difficulty of providing a naturalistic account of such events that leads many critics to deny that they ever occurred; though this looks suspiciously like begging the question in favour of naturalism. It is clear that if such events have occurred, the advance of science has made them more, rather than less, difficult to explain in terms of natural causes. Employing a "God of the gaps" argument that the occurrence of such events would constitute good evidence for supernatural intervention within the natural order seems entirely legitimate.[56]

So, I would maintain that there are gaps and then there are gaps. It's not just pointing to an unsolved engineering problem in nature. The categories of the mental and the physical are logically incompatible categories. You start attributing mental properties to physics and you might end up being told that you are no longer describing the physical at all. Purpose, normativity, intentionality or about-ness—all these things are not supposed to be brought in to the physical descriptions of things, at least at the most basic level of analysis.

Consider the gap between the propositional content of thought and the physical description of the brain. My claim is that no matter how throughly you describe the physical state of the brain (and the environment), the propositional content of thought will invariably be undetermined. This isn't my claim or C. S. Lewis's; this argument was made by the arch-naturalist W. V. Quine. As I see it, it's not a matter of getting a physical description that will work. The logico-conceptual gap is always going to be there regardless of how extensively you describe the physical. As I said earlier, bridging the chasm isn't going to simply be a matter of exploring the territory on one side of the chasm.

I conclude, therefore, that "God of the gaps" or even a "soul of the gaps" response to the argument from reason does not work. I am not saying that we just cannot figure out right now why the mental states involved in rational inference are really physical. I am suggesting on principled grounds that a careful reflection on the nature of mind and matter will invariably reveal that there is a logical gap between them that in principle can't be bridged without fudging categories.

Endnotes

1. Victor Reppert, *C. S. Lewis's Dangerous Idea* (Downer's Grove: Inter-Varsity Press, 2003).
2. C. S. Lewis, *Miracles: A Preliminary Study*, rev. ed. (New York: Macmillan, 1978), 15.
3. Ibid., 213.
4. These features of a broadly materialist position were first developed in William Hasker's *The Emergent Self*. (Ithaca: Cornell University Press, 1999), 59–64.
5. C. S. Lewis, *Surprised by Joy* (San Diego: Harcourt Brace, 1955).
6. C. S. Lewis, "The Empty Universe," in W. Hooper, editor, *Present Concerns* (San Diego,: Harcourt Brace, 1986), 81–82.
7. Daniel Dennett, "Why the Law of Effect Will Not Go Away," *Journal for the Theory of Social Behavior* 5, no. 2 (1976): 171.
8. Susan Blackmore, *The Meme Machine* (Oxford: Oxford University Press, 1999), 236.
9. Robert Van Gulick, "Consciousness," *Stanford Encyclopedia of Philosophy*, accessed May 2, 2012, http://plato.stanford.edu/ spr2007/entries/consciousness.
10. John Beversluis, *C. S. Lewis and the Search for Rational Religion* (Amherst, NY: Prometheus Books, 2007), 148–149.
11. Lewis, *Miracles* (1978), 14.
12. Henry Allison, "Kant's Refutation of Materialism," *The Monist* 79 (1989): 190–209.
13. Arthur James Balfour, *The Foundations of Belief* (London: Longmans, Green, and Co., 1895).
14. The copy of *The Foundations of Belief* that Lewis inherited from his father resides in the collection of the Wade Center at Wheaton College.
15. Arthur James Balfour, *Theism and Humanism* (London: Hodder and Stoughton, 1915).
16. See C. S. Lewis, "Is Theology Poetry?" in *The Weight of Glory and Other Addresses*, rev. and expanded edition, edited by Walter Hooper (New York: Macmillan, 1980), 77; and *The Christian Century*, June 6, 1962.
17. J. B. Pratt, *Matter and Spirit* (New York: Macmillan Books, 1922).
18. Lewis, *Miracles: A Preliminary Study* (New York: Macmillan, 1947), 26–31,
19. G. E. M. Anscombe, *The Collected Papers of G. E. M. Anscombe: Vol.2: Metaphysics and the Philosophy of Mind* (Minneapolis: University of Minnesota Press, 1981), 224–225.

20. Victor Reppert, "The Lewis-Anscombe Controversy: A Discussion of the Issues," *Christian Scholar's Review* (Fall, 1989): 32–48, and *C. S. Lewis's Dangerous Idea*. I owe this point to Jim Slagle.
21. C. S. Lewis, *The Abolition of Man* (New York: Macmillan, 1955), 30.
22. "Irrational," accessed May 12, 2012, http://dictionary.reference.com/browse/irrational.
23. Anscombe, *Collected Papers*, 226.
24. William Hasker, "What about a Sensible Naturalism: A Response to Victor Reppert," *Philosophia Christi* 5 (2003): 61.
25. Ibid.
26. Reppert, *C. S. Lewis's Dangerous Idea*, 73.
27. Anscombe, *Collected Papers*, 226–229.
28. Lewis, *Miracles* (1978), 19.
29. Ibid.
30. Ibid.
31. Ibid., 18.
32. Ibid.
33. Anscombe, *Collected Papers*, 229.
34. Keith Parsons, "Need Reasons be Causes: A Further Reply to Victor Reppert's Argument from Reason," *Philosophia Christi* 5, no. 1 (2003): 72.
35. Lewis, *Miracles* (1978), 16.
36. C. S. Lewis, *Mere Christianity* (New York: Macmillan, 1960), book 3, chapter 11.
37. Paul Draper, quoted on the Internet Infidels home page, accessed May 11, 2012, http://www.infidels.org/infidels/.
38. Anscombe, *Collected Papers*, ix–x.
39. Lewis, *Miracles* (1978), 19.
40. Ibid., 20.
41. Ibid., 22–23.
42. Eric Mascall, *Christian Theology and Natural Science* (New York: Longmans, 1956), 214–216.
43. J. R. Lucas, *Freedom of the Will* (Oxford: Oxford University Press, 1970).
44. William Hasker, "Transcendental Refutation of Determinism," *Southern Journal of Philosophy* 9 (1973): 175–183.
45. John Beversluis, *C. S. Lewis and the Search for Rational Religion* (Grand Rapids: Wm. B. Eerdmans, 1985). He later thoroughly revised the book, publishing it under the same title with Prometheus Books in 2007.
46. Victor Reppert, "Physical Causes and Rational Belief: A Problem for Materialism?" (Ph.D. Dissertation, University of Illinois at Urbana-Champaign, 1989).
47. Alvin Plantinga, *Warrant and Proper Function* (Oxford: Oxford University Press, 1993) and *Warranted Christian Belief* (Oxford: Oxford University Press, 2000).
48. William Hasker, *The Emergent Self* (Ithica, NY: Cornell University Press, 1999), ch. 3.

49. Reppert, "The Argument from Reason," *Philo* 2, no. 1 (1999): 33–45, 76–89; "Reply to Parsons and Lippard on the Argument from Reason," *Philo* 3, no. 1 (2000): 76–89 "Causal Closure, Mechanism, and Rational Inference," *Philosophia Christi* 3, no. 2 (2001): 473–484; "Several Formulations of the Argument from Reason," *Philosophia Christi* 5, no. 1 (2003): 9–34; "Some Supernatural Reasons Why My Critics are Wrong: a Reply to Drange, Parsons, and Hasker," *Philosophia Christi* 5, no. 1 (2003): 77–92.

50. Reppert, *C. S. Lewis's Dangerous Idea*; "The Argument from Reason and Hume's Legacy," in J. Sennett and D. Groothuis, editors, *In Defense of Natural Theology* (Downer's Grove: Inter-Varsity Press, 2005); "Miracles: C. S. Lewis's Critique of Naturalism" in B. L. Edwards, editor, *C. S. Lewis: Life, Works, and Legacy* (New York: Praeger, 2007), vol. 3, 153–82.

51. Darek Barefoot, "A Response to Richard Carrier's Review of *C. S. Lewis's Dangerous Idea* (2007)," http://www.infidels.org/library/modern/darek_barefoot/dangerous.html, accessed May 11, 2012.

52. Angus Menuge, *Agents Under Fire: Materialism and the Rationality of Science* (Lanham, MD: Rowman and Littlefield, 2004).

53. Michael Rea, *World Without Design: The Ontological Consequences of Naturalism* (Oxford: Oxford University Press, 2002).

54. C. S. Lewis, "Why I am Not a Pacifist" in *The Weight of Glory and Other Essays* (New York: Macmillan, 1962), 34.

55. William Hasker, *Metaphysics* (Downer's Grove: Inter-Varsity Press, 1983) 49.

56. Robert Larmer, "Is There Anything Wrong with God of the Gaps Reasoning?" *International Journal for Philosophy of Religion* 52 (2002): 129–142.

SOCIETY

[T]he new oligarchy... must increasingly rely on the advice of scientists, till in the end the politicians proper become merely the scientists' puppets.

—C. S. LEWIS, "WILLING SLAVES OF THE WELFARE STATE"

10

C. S. Lewis and the Advent of the Posthuman

James A. Herrick

"[Professor Filostrato:] 'The work is more important than you can yet understand. You will see.'"
— C. S. Lewis, *That Hideous Strength*[1]

"We are at the start of something quite new in the scheme of things."
—Hans Moravec, *Mind Children*[2]

Professor Julian Savulescu is the head of the Uehiro Center for Practical Ethics at Oxford University and a leading proponent of human enhancement, the school of thought that promotes the progressive use of biotechnologies to improve human intellect, moral reasoning, and other traits such as physical strength. Savulescu has argued that deep moral flaws and destructive behaviors point indisputably to the need to employ technology and education to change human nature; either we take this path or we face extinction as a species.

In Savulescu's view, rapidly advancing brain science will provide some of the data necessary to shaping a better human race: "Once we understand the basis of human brain development, we will be able to augment normal brain development in a way that couldn't naturally occur." But smarter people are not necessarily better people, and so another key to better people is found in a deeper understanding of human biology. "[T]here is reason to believe that even aggression is something that can be understood in terms of its biological underpinnings." A clue to human aggression is discovered in "a mutation in the monoamine oxidase A gene," which in the presence of "early social deprivation" has been linked to "anti-social behavior" in at least one study. Savulescu also notes that "[w]e've been able to manipulate human moral behavior and cooperation through the administration of drugs," Prozac providing one prominent example. Other drugs have been shown "to promote trust and willingness to take risks and recovery of trust after betrayal."

According to Savulescu, genetic science, improved pharmaceuticals, and moral education may hasten the emergence of a new and better human race. However, more is needed, including worldwide cooperation "in a way that humans have never so far cooperated." We live in dangerous times, and greater dangers lie ahead. Weapons technology makes possible the annihilation of millions of people. At the same time "liberal democracy" fails to promote "any particular set of values or particular moral education" as it seeks to guarantee "maximum freedom." Why are personal freedoms a risk factor? The answer is found in a condition theologians might call fallen human nature: "We have a human nature that is severely limited in terms of its origins and in terms of its capacity to respond to these new challenges." Human nature thus requires restraint, modification, or both. Our predicament is deep and complex: weapons of mass destruction, a fragmented political scene, excessive devotion to individual freedoms, and an unreliable nature.

Only aggressive research aimed at helping us to "understand our moral limitations and the ways to overcome these" will produce the scientifically grounded ethics needed to "decide how we should reshape our nature." Employing a vivid analogy, Savulescu affirms that Western culture is entering a dangerous "Bermuda Triangle with liberal democracy in the position of Miami, radical technological advance in the position of Bermuda, and human nature and its limitations in the position of Puerto Rico." To "avoid entering this triangle" will mean reducing "one of our commitments to these points." It is neither likely nor desirable that we would restrain technology, so Bermuda remains on the map. Savulescu continues:

We could reduce our commitment to liberalism. We will, I believe, need to relax our commitment to maximum protection of privacy. We're already seeing an increase in the surveillance of individuals, and that surveillance will be necessary if we're to avert the threats that those with anti-social personality disorders, psychopathic personality disorders, fanaticism represent through their access to radically enhanced technology.

So, liberal Miami is threatened and the dubious Puerto Rico of human nature is clearly targeted for radical change. Of the three points of the Savulescu Triangle only technological Bermuda is safe, a contemporary manifestation of Francis Bacon's island of Bensalem.

Moral education founded on a new ethics is critical to the task of rescuing lost humanity. "I believe that we should be promoting certain sets of values and engaging in moral education," says Savulescu. Tacking away from Miami will require reducing consumerism and accepting "a lower standard of living." Political and economic austerity are also necessary. "We'll need to accept an ethics of restraint, and we'll need to adopt long-term strategies that go beyond a typical electoral term of three to five years." What Savulescu terms "the very extreme adherence to liberalism that we've so far enjoyed" may also have to go.

Of course, human nature will not quickly abandon a comfortable life for a new austerity. After all, we possess "a set of dispositions that make us very ill-disposed to give up our standard of living, to collectively cooperate to solve the world's global problems." Ultimate answers "may lie not only in terms of our political institutions and the degree to which we curb our commitment to liberalism, but also inside ourselves." But, help is at hand because Bermuda survives: "The genetic and scientific revolution that we're a part of today represents a second great human enlightenment." We now possess the means of understanding "the human condition," and we are moving toward an "understanding of our nature as animals, of our dispositions to act, why some people will kill, why some people will give."

Stopping at nothing, "we should adopt whatever strategies are most effective at protecting our future," which includes "moral education, the inculcation of various values and ways of living are no doubt an important part of this." But the greatest obstacle to our survival and advancement is human nature itself; but it must be changed. As impossible as such a transformation might seem to a layperson, Savulescu is hopeful. "[I]t may be that as science progresses, we have at our fingertips the ability to change our nature." The power to transform humanity at the genetic level is in our hands, but "it's up to us to make a decision whether we'll use that power."[3]

This chapter compares certain warnings in C. S. Lewis's *The Abolition of Man* (1944) with recent arguments about our obligation to deploy biotechnologies to alter or "enhance" the human race. I begin with Julian Savulescu because he articulates clearly the values of a growing scientific and cultural phenomenon known as the human enhancement movement or Transhumanism. Not at this point a coordinated effort, human enhancement nonetheless represents the convergence of powerful cultural narratives, mind-boggling technological developments, and a progressive agenda with an improved humanity as its focus. Savulescu's comments serve as an entry point for familiarizing ourselves with the goals and the reasoning of the human enhancement movement.

In order to understand Lewis's objectives in *The Abolition of Man*, particularly the most commented-upon third lecture from which the book takes its title, it will be important to set the work in its historical context. To what specific threats was the great Christian apologist responding in the 1940s? Answering this question makes clear that Savulescu and other enhancement proponents did not invent the agenda they advocate. Today's proponents of biotechnological and ethical improvements to the human race write in a tradition that includes such intellectual luminaries as the eugenics theorist Francis Galton, playwright George Bernard Shaw, scientists such as J. B. S. Haldane and J. D. Bernal, and science fiction writers H. G. Wells and Olaf Stapledon, all figures with whom Lewis was quite familiar. A crucial historical development, however, separates today's advocates from their intellectual predecessors and from C. S. Lewis. No longer are technological alterations to the human constitution a matter of speculation only; they are now vigorously promoted scientific realities awaiting the political and cultural conditions that will allow their implementation.

In an almost uncanny fashion Savulescu's comments reflect key elements of the educational, ethical, and scientific planning that Lewis was concerned to answer in *The Abolition of Man* as well as in his fictional work, *That Hideous Strength* (1945). Proposals by Savulescu and others who share his concerns thus provide an ideal opportunity for assessing the prophetic nature of Lewis's concerns about applied technology in the context of an ascendant Western science operating outside the limits of widespread traditional values Lewis dubbed the *Tao*.

Lewis employed the term *scientism* when discussing science characterized by principles and practices tending toward controlling rather than investigating nature. Science joined to modern ideologies also encouraged the kind of centralized planning he targets in *The Abolition of Man* and elsewhere. Finally, this pivotal distinction between science and scientism, his derisive fictional portrayals of some—though not all—scientists, and provocative comments in letters and essays all raise the question of Lewis's attitude toward science and scientists. Examining *The Abolition of Man* in its historical context will provide help in answering this persistent question.

Lewis's arguments regarding technological modifications to human nature merit attention—even urgent attention—in an era in which human genetic structure may soon be shaped according to the moral vision of a relatively small group of decision-makers. Moreover, his suspicion of scientific planning cut free from traditional values needs to be understood in an age in which technology is advancing at an exponential rate while moral knowledge in the West is declining almost as precipitously.

The Abolition of Man

PROFOUND CONCERNS ABOUT THE DIRECTION of Western education and science prompted C. S. Lewis to pen the three lectures that were first published as *The Abolition of Man* in 1944. In the third talk Lewis argued famously that the power to affect the entire subsequent history of the human race will be determined by a few technologists and bureaucratic planners who alter foundational components in human biology.

The three brief chapters making up *The Abolition of Man*—"Men without Chests," "The Way," and "The Abolition of Man"—were originally presented as the Riddell Memorial Lectures at the University of Newcastle in February of 1943. In the most discussed lecture and the one from which the book takes its title, Lewis warns that "if any one age really attains, by eugenics and scientific education, the power to make its descendants what it pleases, all men who live after it are the patients of that power." Far from being freer and better humans, these new creatures will be "weaker, not stronger: for though we may have put wonderful machines in their hands we have preordained how they are to use them."[4] As a result of sophisticated biotechnology, "Man's conquest of Nature, if the dreams of some scientific planners are realized, means the rule of a few hundreds of men over billions upon billions of men." Lewis concludes: "The final stage is come when Man by eugenics, by pre-natal conditioning, and by education and propaganda based on a perfect applied psychology, has obtained full control over himself. *Human* nature will be the last part of Nature to surrender to Man. The battle will have been won."[5]

Lewis's deep suspicion of modernist educational projects, subjectivism about morality, and progressive scientific planning animates these lectures. He was particularly concerned about biotechnological experimentation with humanity as its "patient," a possibility he also explored in *That Hideous Strength*. In both works Lewis casts a dark vision of the human race redesigned by scientific programmers who have stepped outside the *Tao*.[6] Should such a project succeed, every individual human being eventually would reflect in her or his very cells a new nature crafted by technologists. The new human nature would mirror a moral vision founded on popular but largely unexamined mythologies such as progress and evolution, narratives shaping even scientific planning. Lewis wrote famously, "For the power of Man to make himself what he pleases will be the power of some men to make other men what *they* please."[7] These "man-moulders of the new age will be armed with the powers of an omnicompetent state and an irrepressible scientific technique: we shall get a race of conditioners who really can cut out all posterity in any shape they please."[8]

For the Conditioners—Lewis's label for the scientists and bureaucrats at work on the new humanity—values and emotions are mere physical phenomena to be produced or repressed in students through education informed by advanced psychology. Breaking with past traditions, values become an educational outcome to be propagated rather than a deeply rooted moral awareness to be refined, "the product, not the motive, of education." The Conditioners will acquire the capacity to "*produce* conscience and decide what kind of conscience they will produce."[9]

Future academic and governmental elites will define "good" and then set about producing this invented good in humankind by a combination of educational technique and biotechnology. According to Lewis, they "know quite well how to produce a dozen different conceptions of good in us," though guided by no external, objective standard of good themselves. Ignoring the timeless *Tao*, the Conditioners become the arbiters of good and bad.[10] But, for Lewis, to step outside the *Tao* is to sever one's moral connection with all previous human experience, in essence, to cease to be human. To propagate this moral rupture by technological means is to create, not an improved human race, but a race that is no longer human; this is the abolition of man.

Lewis, Science and Science Fiction

It would be easy to read the third lecture of *The Abolition of Man* as an attack on science, a view that appears to receive support from other works by Lewis. But, was Lewis an enemy of science? The apparent answer to this question is no, for he followed scientific developments and spoke respectfully even of scientists with whom he disagreed, such as J. B. S. Haldane. Lewis was curious about and sought to grasp contemporary scientific theories, and was willing to embrace what he took to be legitimate scientific insights. He corresponded for a time with Sir Arthur C. Clarke, himself a scientist. Lewis also fashioned the tough-minded physical chemist Bill Hingest in *That Hideous Strength* to represent a "real scientist." Bill "The Blizzard" has no patience with science as modern magic or social programming, a position for which his colleagues at the N.I.C.E. (National Institute for Coordinated Experiments) label him a reactionary, and for which he is murdered.[11]

In short, Lewis respected scientific work that pursued knowledge of the natural world. Science, for Lewis, was a means of *seeing* and thus of appreciating nature, and the inviolability of nature was the principal value guiding its investigations. By contrast, the new science, or scientism, developed out of an impulse to *see through* nature by deconstructing its processes until everything in it—including the human being—was explained as a matter of mere physical causality. Scientism's ultimate goal is placing all of nature under human control.

Because he feared the rising influence of scientism, Lewis maintained a deep suspicion of some scientists, some scientific projects, and all centralized scientific planning. These suspicions are evident in several of his essays and letters, and in his three science fiction novels. The villainous genius in the first two of these books is a diabolical scientist with the suggestive name, Weston, while a cabal that includes scientists works its devilry in *That Hideous Strength*. Weston exports murder and other sins to Mars in *Out of the Silent Planet* (1938), and becomes the devil incarnate in *Perelandra* (1943), seeking to provoke a human fall from grace on Venus. These are not subtle literary gestures on Lewis's part, and each targets scientists and scientific programs.

In a letter to Clarke written in 1943—the year he delivered the lectures that became *The Abolition of Man*—Lewis makes clear his antipathy toward the hubris-driven scientific agenda he saw emerging in the West:

> I don't of course think that at any moment many scientists are budding Westons: but I do think (hang it all, I *live* among scientists!) that a point of view not unlike Weston's is on the way. Look at Stapledon (*Star Gazer* [sic] ends in sheer devil worship), Haldane's *Possible Worlds* and Waddington's *Science & Ethics*. I agree Technology is *per se* neutral: but a race devoted to the increase of it[s] own forces & technology with complete indifference to either *does* seem to me a cancer in the universe. Certainly if he goes on his present course much further man can *not* be trusted with knowledge.[12]

Lewis's reference to "a cancer in the universe" seems aimed at the human race rather than the scientific caste that he says he lives among. But Lewis perceived a link between the two, a bridge spanning the gulf separating academic science and the general public. Again, it was not scientific research that concerned Lewis but modernist philosophical convictions scientists had adopted to guide their work, notions such as naturalism, evolutionism (evolution as incremental improvement), and technological determinism. Of even greater concern to Lewis was the ethical error of thinking that objective moral principles can be discovered in an absolute external to the *Tao*. To pursue such a possibility was to place oneself outside the universal *Tao*, to sever one's connection to human moral history, and thus to cease to be a fully human.

Lewis was not principally concerned to answer other academic figures in his polemical writing. Unlike most intellectual leaders of his day, he was acutely aware of the power of popular narrative and mass media to mold public conceptions of science and even to shape political policy. His unlikely entry into the science fiction arena was an effort to address, in a genre slighted by literary scholars, ideas he considered dangerous enough to attack by direct argument in *The Abolition of Man*. Ongoing human evolution and the inevitable conquest of space—two popular ideas that concerned Lewis deeply—had been advocated for decades in engaging novels by H. G. Wells and Olaf Stapledon. Lewis perceived a spiritual threat in such entertainments, a tendency toward heresy following the avenue of the imagination. Not only did science fiction promote dubious technological goals such as propelling a fallen human race into space, but it also suggested omnipotence and omniscience outside of God, redemptive prospects other than the Crucifixion, and a future determined by human motive rather than divine will.

According to Lewis, the greatest threat to public attitudes was the mistaken view, often implicit in science fiction, that moral absolutes could be discovered within the scientific enterprise itself. The science fiction with which Lewis was familiar portrayed in vivid narrative a Western scientific project that was ethically self-justifying. To answer such errors would require fashioning similarly compelling narratives in which scientism was exposed as a dangerous moral fraud, and in which transcendent values external to science were ultimately triumphant. Lewis's three works of science fiction represent a sustained and perhaps unprecedented effort by a major intellectual figure to participate in a public controversy by answering opponents in their chosen medium of popular fiction.

The Abolition of Man articulates the intellectual basis for Lewis's fictional assault on dogmatic science: Over all human enterprise across the millennia stands the moral unity of the *Tao*, a timeless and universal expression of value reflecting the moral nature of God himself. The *Tao* is not exclusively expressed in a single spiritual tradition, but reflected in all enduring religious and philosophical traditions. Each generation bequeaths the *Tao* to the next; we might ignore it or seek to reason it out of existence, but we are not free to alter it. To seek to operate outside of its influence is to retreat to fallible reason following dangerous ideologies. Lewis warns readers of a day in the near future when institutionalized science, ignoring the *Tao*, decides on the basis of the dangerous moral presumption that it is preserving humanity, to change human nature.

Lewis's treatment of the moral devolution of science is carefully nuanced. The impulse to preserve and propagate the human race is not itself evil, but morally insufficient, an enduring fragment from a fractured memory of the *Tao*. In *Out of the Silent Planet* the Oyarsa or governing angel of Malacandra tells Weston that the one component of the *Tao* the devil—or Bent One—has allowed him to revere "is the love of kindred. He has taught you to break all of them except this one."[13] Extending the presence and influence of the human race without regard for other complementary and restraining moral precepts is the catastrophic error of the scientist Weston, and of the Western scientism he represents.

A number of writers of his own and the immediately previous generation contributed to Lewis's concerns about this "cancer in the universe," the forceful extension of human influence without regard for nature or individual rights. In the twentieth century's opening decades the rising yet critically maligned genre of science fiction provided a narrative laboratory for testing ideas such as space travel and continuing human evolution. H. G. Wells (1866–1946), a prolific and popular writer, contributed more than any other figure to the vision of the future taking shape in the public mind. Space exploration, human evolution, technological progress, and dangerous alien life-forms were prominent themes; conquest of the universe was human destiny. "But for Man, no rest and no ending," wrote Wells in his 1933 novel, *The Shape of Things to Come*. Evolution and determination will carry us as a conquering force out into the cosmos:

> He [mankind] must go on, conquest beyond conquest. First this little planet with its winds and ways, and then all the laws of mind and matter that restrain him. Then the planets about him and at last out across immensity to the stars. And when he has conquered all the deeps of space and all the mysteries of time, still he will be beginning.[14]

Wells had a deep connection to turn-of-the-century evolutionary science, having studied as a youth at T. H. Huxley's school in London where Darwin himself—a close associate of Huxley—was an occasional visitor.

The great playwright G. B. Shaw (1856–1950) proposed state-sanctioned eugenics programs and advocated radical life extension. In the five plays constituting the *Back to Methuselah* series (1921) Shaw imagined a superior human race to follow the present one, a product of eugenics and technological advancements. Lewis mentions *Back to Methuselah* as reflecting an objectionable view of biological progress in which the individual was expendable in pursuit of a new race. Contemporary Christian writer and friend of Shaw, G. K. Chesterton (1874–1936), wrote: "If man, as we know him, is incapable of the philosophy of progress, Mr. Shaw asks, not for a new kind of philosophy, but for a new kind of man."[15]

Of more urgent concern to Lewis, however, was the philosopher turned science fiction writer, Olaf Stapledon (1886-1950). A legendary figure in the history of the genre, Stapledon made the future-human a central theme of his fiction.[16] *Last and First Men* (1930) and *Star Maker* (1937) projected human evolution into the distant future. Stapledon also maintained an earnest interest in eugenics and deep admiration for Galton. "Columbus found a new world; but Francis Galton found a new humanity," he wrote.[17] Galton's eugenics narrative appealed to Stapledon: "In time it may be as possible to breed good men as it is possible to breed fast horses."[18]

Other statements by Stapledon undoubtedly attracted Lewis's attention. "Darwin showed that man is the result of evolution. Others have shown that he may direct his evolution." Suggesting the religious nature of his commitment to evolution, Stapledon posed the provocative question: "Why is it sacrilegious to use direct means for the improvement of the human race?" Human beings "have been given wherewithal to climb a little nearer to divinity."[19] Stapledon influenced writers of the next generation, including Clarke, whose classic *Childhood's End* (1953) imagines an evolving, collective human consciousness merging into a divine Overmind. Despite his deep misgivings about Stapledon's philosophy, Lewis—always willing to acknowledge true talent—referred to him as an author he could read "with delight."[20]

Wells, Shaw and Stapledon wrote popular narratives reflecting the emerging scientific mythologies of evolving humanity and limitless technological progress.[21] But actual scientists also concerned Lewis, as is clearly evident in the sinister character of Weston in both *Out of the Silent Planet* and *Perelandra*. Lewis was familiar, for instance, with the visionary works of famed Irish scientist J. D. Bernal, a proponent of governmental scientific planning, space exploration, and human enhancement technologies. In books such as *The World, the Flesh and the Devil* (1929) and *The Social Function of Science* (1939) Bernal brought his ideas to the public. Typifying for Lewis the emerging Western scientistic value system, Weston may be based on Bernal. One scholar has argued that Lewis satirized Bernal and his proposal for centralized tracking of scientific research in *That Hideous Strength* as well.[22]

Bernal possessed a flair for provocative prose and a willingness to shock his readers with outlandish narratives of the future. In *The World, the Flesh and the Devil* he affirms that the brain is the essentially human part of the person, an idea Lewis skewers with gruesome realism in *That Hideous Strength*. Bernal approached self-parody in his more extreme proposals:

> After all it is brain that counts, and to have a brain suffused by fresh and correctly prescribed blood is to be alive—to think. The experiment is not impossible; it has already been performed on a dog and that is three-quarters of the way towards achieving it with a human subject.

Such a preserved brain would want to communicate with other persons, however, for "[p]ermanently to break off all communications with the world is as good as to be dead." Science held the solution to even this problem:

> [T]he channels of communication are ready to hand. Already we know the essential electrical nature of nerve impulses; it is a matter of delicate surgery to attach nerves permanently to apparatus which will either send messages to the nerves or receive them. And the brain thus connected up continues an existence, purely mental and with very different delights from those of the body, but even now perhaps preferable to complete extinction.

An early enhancement advocate, Bernal envisioned other alterations to the body and mind. In one passage he provided readers something of a shopping list of posthuman qualities:

> We badly need a small sense organ for detecting wireless frequencies, eyes for infra-red, ultra-violet and X-rays, ears for supersonics, detectors of high and low temperatures, of electrical potential and current, and chemical organs of many kinds. We may perhaps be able to train a great number of hot and cold and pain receiving nerves to take over these functions; on the motor side we shall soon be, if we are not already, obliged to control mechanisms for which two hands and feet are an entirely inadequate number; and, apart from that, the direction of mechanism by pure volition would enormously simplify its operation.[23]

Bernal was not alone in advocating radical biotechnological alterations to the human being. Though no longer a widely recognized name, J. B. S. Haldane (1892-1964)—identified as a threat in a Lewis letter to Clarke—was another popular scientific writer of whom Lewis harbored profound suspicion. Like Bernal, Haldane preached space colonization and enhanced evolution.[24] In *Possible Worlds* (1928) he envisioned humanity evolving into "a super-organism with no limits to its possible progress." There was "no theoretical limit to man's material progress but the subjugation to complete conscious control of every atom and every quantum of radiation in the universe."[25] Civilization's technological advancement was a religious conviction for Haldane.

This dispute between Haldane and Lewis took a personal turn when Haldane responded in print to what he took to be Lewis's attack on science and scientists. Lewis wrote a reply to these criticisms which amounted largely to arguing that Haldane misconstrued the argument of *The Abolition of Man* and of his science fiction. Lewis affirms that "'scientists' as such are not the target." The real problem is "philosophical, not scientific at all" for, as Ransom says in *Out of the Silent Planet*, "the sciences are 'good and innocent in themselves'... though evil 'scientism' is creeping into them."[26] Thus Lewis can write that "under modern conditions any effective invitation to Hell will certainly appear in the guise of scientific planning."²⁷ Insisting that under such circumstances "devil worship is a real possibility," Lewis adds, "I believe that no man or group of men is good enough to be trusted with uncontrolled power over others."²⁸

When an ideology acquires religious stature, a social order taking shape around it will be inclined toward evil. Theocracy of any type was thus for Lewis "the worst" form of government. "A metaphysic, held by the rulers with the force of a religion is a bad sign" because such an outlook "forbids wholesome doubt." And, as a political perspective can never be perfect, religious rigidity is particularly dangerous in governmental planning. A secretive party with access to power and religious certainty about its cause is highly dangerous.²⁹ Add to political power fervent belief in an inevitable force—technology or evolution—and the rule of law itself is at risk. Under such conditions "revolutionary methods" will be justified, for "necessity knows no law."³⁰

But Lewis, author of *The Screwtape Letters* (1942), discerned deeper and more sinister springs of human evil. As already noted, he stated in a letter to Clarke that Stapledon's *Star Maker* ends in "devil worship." Demonic influence is associated with science or scientists in all three of Lewis's works of science fiction. If, as he claims, the phrase "devil worship" does not usually mean that someone "knowingly worship[s] the devil," then to what is Lewis referring with his repeated references to devilry? And, what does this allegation have to do with science? When a group comes to venerate "as God" an idealized state of affairs whose perfect instantiation would result in evil—disregard for the individual, the violation of nature, destruction of the *Tao*—then the accusation of devil worship is warranted. Lewis refers to this eventuality as worshipping one's own vices. "It is clearly in that sense, and that sense only, that my Frost [in *That Hideous Strength*] worships devils." For Lewis, the scientist Frost symbolizes "the point at which certain lines of tendency already observable will meet if produced."³¹

Lewis identified two such tendencies before his unfinished reply to Haldane breaks off. The first such line, writes Lewis, "is the growing exaltation of the collective and the growing indifference to persons... [T]he general character of modern life with its huge impersonal organizations may be more potent than any philosophy." Clearly, Lewis viewed anonymous bureaucracies with suspicion, a conviction expressed in the stinging satire of *The Screwtape Letters* where devils inhabit an elaborately organized and labyrinthine "lowerarchy." Moreover, reverence for the collective was corrosive of individual rights; when "the individual does not matter... we really get going" and "it will not matter what you do to an individual."[32] Lewis's foundational concern is the tendency of collectives toward "the abolition of persons." When the cure for selfishness is, as Haldane had suggested, removing from language words such as "my" or "I," then the elimination of individuals has been brought a little closer.

The second tendency is "the belief of [an organization's] members that they are not merely trying to carry out a programme but are obeying an impersonal force," particularly one to which members attribute inevitability. "The belief that the process which the Party embodies is inevitable, and the belief that the forwarding of this process is the supreme duty and abrogates all ordinary moral laws."[33] Evolution, or the gradualist rendition of Darwinism Lewis referred to as evolutionism and developmentalism, was just such an inevitable force. So pervasive was developmentalist philosophy that "evolution" had become an account of any gradual change over time, and incremental change over time always meant improvement.[34] Scientism adopted evolutionism as the cosmos's own narrative, an account of everything that we experience, and the story of our inevitable future.[35]

Contemporary Transhumanism

LEWIS'S PROPHETIC APPRAISAL OF CERTAIN scientific trends in *The Abolition of Man* finds confirmation in today's discourse of our biotechnological future. The vision of technologically enhanced posthumanity arises out of a synthesis of scientific culture's most robust mythologies— progress, evolution, the superman, and the power of collective intellect. Technology will conquer death, space, and human nature, and deliver us into the future as highly evolved demigods. The Internet is humanity's first major step toward a unified web of consciousness—Teilhard de Chardin's *noosphere*—that will first blanket the earth and then pervade the universe.[36] The objections of "bio-conservatives" will be silenced through popular argument and public art, and the way opened to unlimited progress, miraculous technologies, and visionary ethics. Then comes posthumanity and Bertrand Russell's "world of shining beauty and transcendent glory."[37] Transhumanism affirms that the time has arrived to make good on such prophecies by crafting a technologically enhanced, globally connected and immortal race—Stapledon's "splendid race."

Contemporary Transhumanism draws inspiration from Utopianism, Renaissance Humanism, Enlightenment Rationalism, nineteenth-century Russian Cosmism, New Age Gnosticism, science fiction, speculative techno-futurism, and apocalyptic themes in the Judeo-Christian tradition. Nick Bostrom, Oxford University philosopher and one of the founders of contemporary Transhumanism, captures the movement's fundamental orientation:

> Transhumanists view human nature as a work-in-progress, a half-baked beginning that we can learn to remold in desirable ways. Current humanity need not be the endpoint of evolution. Transhumanists hope that by responsible use of science, technology, and other rational means, we shall eventually manage to become posthuman, beings with vastly greater capacities than present human beings have.[38]

Evolving humanity, long a theme in popular scientific writing and science fiction, has now emerged as a major topic in bioethics, philosophy and religion.[39] Ongoing evolution will "eventually produce a unified cooperative organization of living processes that spans and manages the universe as a whole."[40] Evolution is now a process in which human beings may actively participate by technological means. The present human being is not the crown of evolution's creative work but a step toward something grander—the posthuman. But, even posthumanity is not the ultimate goal. Inexorable evolution is producing, by means of its human and posthuman surrogates, ever more advanced technologies as part of its plan to achieve omniscience and omnipotence. Ambitious evolution is merely using us and our descendants as its cat's paw to snatch technological divinity from the cosmos's chaotic flames.

The specific characteristics of posthumanity are debated; what is crucial is the conviction that the posthumans are near, that they will represent a profound improvement over our present condition, and that we ought to work diligently for their arrival. One Transhumanist advocate affirms:

> Trust in our posthuman potential is the essence of Transhumanism. We trust that we can become posthumans, extrapolating technological trends into futures consistent with contemporary science, and acting pragmatically to hasten opportunities and mitigate risks. We trust that we should become posthumans, embracing a radical humanism that dignifies the ancient and enduring work to overcome and extend our humanity.[41]

The posthuman future is not limited by biology but will involve human beings merging with machines, at first by simply mechanically augmenting the body but eventually by depositing human consciousness in mechanical devices. Thus will we achieve immortality, universal knowledge, and unified global consciousness.

The process of creating posthumanity is fundamentally evolutionary, but with an important difference when contrasted to the old Darwinian model. As Lewis speculated in *The Abolition of Man*, biotechnologies will permit us to be active participants in our own evolution.[42] Transhumanist leader James Hughes writes that we "must accommodate the 'posthumans' that will be created by genetic and cybernetic technologies."[43] "This vision, in broad strokes," affirms Oxford's Bostrom, "is to create the opportunity to live much longer and healthier lives, to enhance our memory and other intellectual faculties, to refine our emotional experiences and increase our subjective sense of well-being, and generally to achieve a greater degree of control over our own lives." According to Bostrom, the aggressive pursuit of biotechnology is a radical reaction against current convention, "an alternative to customary injunctions against playing God, messing with nature, tampering with our human essence, or displaying punishable hubris."[44] Efforts to coax the public to embrace the ideology of posthumanity, however, will surely provoke a contest. Thus, Hughes predicts that "the human race's use of genetic engineering to evolve beyond our current limitations would be a central political issue of the next century."[45]

More may be ahead than domestic political debate, however. According to some experts, the near future will usher in a global culture enabled by a massively more powerful Internet. Computer engineer Hugo de Garis takes as simple matters of fact that "the exponential rate of technical progress will create within 40 years an Internet that is a trillion times faster than today's, a global media, a global education system, a global language, and a globally homogenized culture" which will constitute the basis of "a global democratic state." This new order of things, which de Garis calls Globa, will rid the world of "war, the arms trade, ignorance, and poverty."[46] The coming transformation of the human race and the world it inhabits is nothing short of an apocalypse—the Kingdom arrives via the Internet.

What was "previously sought through magic and mysticism," writes Hughes, will now be pursued technologically.[47] Bostrom imagines a utopia in which posthumans enjoy "aesthetic and contemplative pleasures whose blissfulness vastly exceeds what any human being has yet experienced." The new people will experience "a much greater level of personal development and maturity than current human beings do, because they have the opportunity to live for hundreds or thousands of years with full bodily and psychic vigor." He continues:

> We can conceive of beings that are much smarter than us, that can read books in seconds, that are much more brilliant philosophers than we are, that can create artworks, which, even if we could understand them only on the most superficial level, would strike us as wonderful masterpieces. We can imagine love that is stronger, purer, and more secure than any human being has yet harbored.[48]

Bostrom and Hughes strike a winsome note in their predictions of the posthuman future. However, at what cost does the New Era arrive? Will we forgo individual rights, as Lewis feared, in the pursuit of a greater collective good? Science writer Ronald Bailey contends that democratic majorities often oppose "avant-gardes minorities." If "the transhuman future we are all hoping for" is to be achieved, it may require efforts more aggressive than those suggested by Bostrom's irenic reverie. Regrettably, democracy often has placed limits on cutting-edge scientific research. Bailey argues that in some "benighted jurisdictions" promising research agendas can be stopped in their tracks by "majoritarian tyranny." Despite the apparent lessons of history regarding programs for improving humanity, Bailey looks hopefully toward the day when an emerging posthuman race will transform the world—that is, if democracy doesn't get in the way.[49] Perhaps Lewis's fears about religious devotion to inevitable processes were well-founded.

Considerably more reassuring to wary audiences is the central figure in the contemporary human enhancement movement, inventor Ray Kurzweil, best known for his theory of exponential technological progress culminating in the Singularity. At a moment in time not more than a few decades away, a technological explosion will change everything permanently. Kurzweil's vision of a transformative human future has recently captured public attention in books such as *The Singularity is Near* and movies such as *Transcendent Man*.[50] He confidently affirms that exponential progress in the biological sciences will soon allow us to "reprogram the information processes underlying biology."[51] While the idea here is vague and expressed for a lay audience, the planned reprogramming of foundational human biology is the specific goal of Lewis's Conditioners. For Kurzweil and other techno-futurists, the future will reveal unimaginable improvements to the human condition. Nature will yield to technology; the battle will have been won.

Kurzweil has become the public face of human enhancement, an affable front man with an accountant's demeanor. The heavy theoretical lifting, however, is done by others. Philosopher John Harris, among the four or five leading apologists for human enhancement, argues that assisting evolution is a moral obligation. He writes, "The 'progress of evolution' is unlikely to be achieved accidentally or by letting nature take its course." Joining Savulescu in urging the necessity of enhanced evolution, Harris argues that "if illness and poverty are indeed to become rare misfortunes, this is unlikely to occur by chance… It may be that a nudge or two is needed: nudges that will start the process… of replacing *natural selection* with *deliberate selection, Darwinian evolution* with *'enhancement evolution.'*"[52] While Harris's metaphor suggests a gentle technological push along coordinates of improvement already plotted out by nature, it would be wide of the mark to imagine that science has identified such an evolutionary trajectory for future humanity. It is more likely that educated guesses grounded in hopeful narratives about progress substitute for actual knowledge in this and similar scenarios.

An inevitable force with motives of its own, evolution is central to the techno-futurists' vision of the posthuman future. Evolution produced us and through us, technology. It, not God and not the *Tao*, is also the source of the moral principles that have brought us to the point of transformation as a species, and that will ensure our continued evolution. Computer scientist Hugo de Garis affirms that "because of our intelligence that's evolved over billions of years, we are now on the point of making a major transition away from biology to a new step. You could argue that… maybe humanity is just a stepping stone."[53] Physicist Freeman Dyson agrees—we will be transformed as "many opportunities for experiments in the radical reconstruction of human beings" present themselves.[54] But there is more to our posthuman future than simply improving our lot here on earth: The new humanity, toward which the present human race represents a mere step along the way, will propagate itself throughout the cosmos. This was the cosmic vision of scientific planners and science fiction authors that prompted Lewis's skepticism about space exploration. Sounding a theme reminiscent of Wells, Dyson writes that "when life and industrial activities are spread out over the solar system, there is no compelling reason for growth to stop."[55] Technologically assisted evolutionism is becoming, as Lewis warned, a comprehensive narrative of an inevitable force's ultimate universal triumph.

Human enhancement advocates focus attention on four technologies—nanotechnology, biotechnology, information technology, and cognitive science, or NBIC. But technology is not the whole story of the turn toward Transhumanism. "The NBIC technologies," writes Hughes, will "change how we work, how we travel, how we communicate, how we worship and how we cook."[56] Whereas work, travel, and communication are perhaps expected in this list, and cooking seems trivial by comparison, "how we worship" is arresting. Traditional religion has been the *bête noir* of enhancement advocates, an anti-technological and anti-futurist force to be actively opposed. Hughes's comment, however, hints at a new approach—the re-imagining of religion along Transhumanist lines. For some in the movement posthumanity and advanced technologies are objects of worship, hope in the Singularity a religious faith. The new wine of Singularity religion will require the new wine skins of innovative religious expression; techno-futurism will discover transcendence in techno-religion.

Critical Reaction

Not surprisingly, contemporary Transhumanism has attracted a number of informed critics. I will briefly review two prominent voices in the opposition camp who reflect concerns at the heart of Lewis's own case. Hava Tirosh-Samuelson, a skeptic as regards the Transhumanist vision, echoes one of the central arguments of *The Abolition of Man*—biotechnology now threatens to exercise control of nature itself:

> Due to genetic engineering, humans are now able not only to redesign themselves… but also to redesign future generations, thereby affecting the evolutionary process itself. As a result, a new *posthuman* phase in the evolution of the human species will emerge, in which humans will live longer, will possess new physical and cognitive abilities, and will be liberated from suffering and pain due to aging and diseases. In the posthuman age, humans will no longer be controlled by nature; instead, they will be the controllers of nature.[57]

The question of altering *human* nature also remains at the center of the developing case against Transhumanism and related proposals. Famed historian Francis Fukuyama, for example, has argued that "contemporary biotechnology" raises "the possibility that it will alter human nature and thereby move us into a 'posthuman' stage of history." This possibility poses a real danger to individual rights and threatens the foundation of democratic institutions:

> This is important... because human nature exists, is a meaningful concept, and has provided a stable continuity to our experience as a species. It is, conjointly with religion, what defines our most basic values. Human nature shapes and constrains the possible kinds of political regimes, so a technology powerful enough to reshape what we are will have possibly malign consequences for liberal democracy and the nature of politics itself.[58]

Though the deeper dangers of biotechnological alterations of humans have not yet manifested themselves, Fukuyama adds, "one of the reasons I am not quite so sanguine is that biotechnology, in contrast to many other scientific advances, mixes obvious benefits with subtle harms in one seamless package."[59] The essential correctness of Lewis's case is evident in the duration of major components in his rebuttal to Bernal, Stapledon, Haldane, Shaw and other enhancement proponents of his own day.

C. S. Lewis exhibited remarkable prescience in *The Abolition of Man*. Was there anything that he failed to see? Writing in the war years of the early 1940s, Lewis's perspective was understandably shaped by present circumstance and personal experience. As a result, he did not anticipate certain cultural and historical developments that have become critical to the rise of posthumanity thinking.

As noted, Lewis harbored a deep antipathy for faceless state institutions where atrocities are plotted out according to cost-benefit pragmatism and inhuman schemes are hatched in dingy meeting rooms. In such settings was the banality of evil expressed in war-torn Europe. Lewis does not appear to have anticipated the postwar power of the large corporation, the modern research university, and sophisticated mass media. Such shapers of twenty-first century American culture, not the cumbersome state agencies of mid-century Europe, have taken the lead in developing the biotechnologies, educational techniques, and persuasive prowess Lewis cautioned against. The user-friendly smile of the high-tech firm, not the icy stare of a government department, is the face of the new humanity. Moreover, justifications for enhancement research are not hammered out in centralized planning meetings, but tested on focus groups and winsomely presented in entertaining public lectures. Financial support for posthumanity comes not come from Big Brother bureaucracies but from Silicon Valley boardrooms.

The scope of research related to human enhancement is incomprehensibly vast and accelerating at an incalculable rate. Hundreds and perhaps thousands of university and corporate research facilities around the world are involved in developing artificial intelligence, regenerative medicine, life-extension strategies, and pharmaceutical enhancements of cognitive performance. An ever-increasing number of media products including movies, video games, and novels promote Transhumanist and evolutionist themes. Each technological breakthrough is promoted as a matter of consumerist necessity despite the fact that personal electronic devices—and the companies marketing them—are increasingly intrusive and corrosive of personal freedoms. Innovative educational organizations such as Singularity University are forming around the Transhumanist ideal. Indeed, so immense, diverse, and well-funded is the research network developing enhancement technologies that the collective financial and intellectual clout of all related projects is beyond calculating. Suffice it to say that the enhancement juggernaut is astonishingly large and powerful.

Conclusion

Is there a real threat about which we ought to be concerned in light of *The Abolition of Man*, a text now seventy years old? As we have seen, C. S. Lewis harbored profound reservations about science operating independently of the *Tao* and driven by philosophies that claimed scientific support. He believed that human nature itself was at risk, and that the entire subsequent history of the human race could be adversely affected by a small cadre of scientific and bureaucratic planners, his Conditioners.

This essay has argued that Lewis was prophetic as regards the advent of techniques powerful enough to bring about the effects he feared. I also contend that the scientistic mythologies propelling contemporary calls for unhindered biotechnological research—the mythologies against which Lewis cautions readers in *The Abolition of Man*—are contributing to conditions that encourage potential misuse of enhancement technologies. The remaining question, then, is whether such technology will be used to bring about the ends Lewis feared.

Toward the end of *The Abolition of Man* Lewis argues that modern science requires restraint, and that such restraint is unlikely to come from within science. While no one can question Western science's great accomplishments, "its triumphs may have been too rapid and purchased at too high a price." As a result, Lewis contends that science must be recalled to a position of submission to deeper and older human truths. In a rebuke that sounds almost naïve in a twenty-first century context, Lewis suggests that "reconsideration, and something like repentance, may be required." Only "a regenerate science" would recognize the wisdom in "buy[ing] knowledge at a lower cost than that of life."[60] Lewis considered that the threat of coordinated technology, psychology, and ethics to the future of the human race was real, and "if the scientists themselves cannot arrest the process before it reaches the common Reason and kills that too, then someone else must arrest it."[61]

Will Lewis's prophecy of biotechnology's misuse come to pass? It is my view that if the enhancement juggernaut continues to gain speed, financial power, and cultural force, as it shows every sign of doing, a clash of two essentially religious worldviews is in the offing: Judeo-Christian theism and Transhumanist techno-futurism. Religious traditionalists will not readily accept a biotechnological project aimed at creating new human beings at the expense of human nature, and human enhancement advocates will not yield their vision of a posthuman future to the obscurantist objections of opponents they dismiss as bio-conservatives. If Lewis is right, this will be a battle for the future of the human race itself. I am convinced he is right. Currently, one side in the contest is preparing itself for a long cultural struggle, developing its resources, crafting its arguments, honing its public case. The other side has yet to realize that a battle is in the offing. For too long people of faith have adored Lewis without hearing him, pursued adulation without reflection, allowed his brilliant imagination to entertain without allowing his piercing intellect to chasten. We read Lewis as a friendly uncle; we need to encounter him as a fiery prophet.

C. S. Lewis provides a model for responding in the public sphere to a mistaken understanding of science itself as a source of moral value, a view that moves both science and society dangerously outside the boundaries of the *Tao*. Differentiating between science per se and the *scientism* that seeks to control nature itself, Lewis criticized scientists who functioned as prophetic visionaries, social programmers, and agents of human transformation. In *The Abolition of Man* Lewis reveals the errors in ethical speculation that drove Western science from the *is* of discovery to the *ought* of moral prescription. Technology now advances at a rate more rapid than even the most dedicated observer is capable of tracking. Our contemporary moral guides offer us the astonishing speed of progress as assurance of the unquestionable correctness of progress; *rate* of change now equals *rightness* of change.

Lewis concludes *The Abolition of Man*, curiously, with a discussion of magic and science in the sixteenth and seventeenth centuries. Referring to the two enterprises as "twins" born at the same time and under the same circumstances, he notes that magic eventually died while science grew and flourished. Both, nevertheless, were seeking the same end, and by similar means. The following comparison no doubt angered scientists in his day, as Lewis must have known it would: "For magic and applied science the problem is how to subdue reality to the wishes of men: the solution is a technique; and both, in the practice of this technique, are ready to do things hitherto regarded as disgusting and impious—such as digging up and mutilating the dead."[62] Today we are on the verge of altering the living by manipulating their genetic structure, an activity that would strike Lewis as similarly magical in impulse and similarly wrong in moral intent.

Endnotes

1. C. S. Lewis, *That Hideous Strength* (New York: Macmillan, 1967), 60.
2. Hans Moravec, *Mind Children: The Future of Robot and Human Intelligence* (Cambridge MA: Harvard University Press, 1988). 158.
3. Julian Savulescu, "Unfit for Life: Genetically Enhance Humanity or Face Extinction," keynote address, *Festival of Dangerous Ideas* (Sydney, Australia, 2009).
4. C. S. Lewis, *The Abolition of Man* (New York: Macmillan, 1947), 36.
5. Ibid., 37.
6. Lewis provides readers a formulation of the *Tao* as an appendix to *The Abolition of Man*.
7. Lewis, *Abolition of Man*, 37.
8. Ibid., 38.
9. Ibid., 39.
10. Ibid., 39.
11. Lewis, *That Hideous Strength*, 56–58, 70–72, 77–83,
12. Arthur C. Clarke, C. S. Lewis, *From Narnia to a Space Odyssey: The War of Ideas between Arthur C. Clarke and C. S. Lewis* (New York: iBooks, 2003), 40.
13. C. S. Lewis, *Out of the Silent Planet* (San Francisco, Harper Collins: 2005), 137.
14. H. G. Wells, *The Shape of Things to Come* (London: Penguin Classics, 2005).
15. G. K. Chesterton, *Heretics* (London: John Lane, 1905), 66. Chesterton understood the progressive eugenic view of his day as having taken on a religious quality. "But the sensation connected with Mr. Shaw in recent years has been his sudden development of the religion of the Superman." Ibid, 62.
16. Sam Moskowitz, *Explorers of the Infinite: Shapers of Science Fiction* (Westport, CT: Hyperion Press, 1963), 261.
17. Olaf Stapledon, "The Splendid Race" in *An Olaf Stapledon Reader*, ed. Robert Crossley (Syracuse, NY: Syracuse University Press, 1997), 145, 146.
18. Ibid., 145.
19. Ibid., 147.
20. Lewis, *Abolition of Man*, 25.
21. On the idea of mythologies of science, see my *Scientific Mythologies: How Science and Science Fiction Forge New Religious Beliefs* (Downers Grove, IL: InterVarsity Press, 2008).
22. Dale Sullivan, "C. S. Lewis' Satirical Portrait of J. D. Bernal's *The Social Function of Science*," paper presented at the 95th meeting of the National Communication Association, Chicago, IL (November 2009).

23. J. D. Bernal, *The World, The Flesh and the Devil: An Enquiry into the Future of the Three Enemies of the Rational Soul* (Bloomington, IN: Indiana University Press, 1969). Also available at http://cscs.umich.edu/~crshalizi/Bernal/, accessed May 10, 2012.
24. J. B. S. Haldane, *Possible Worlds* (London: Chatto and Windus, 1937), 301.
25. Ibid.,, 304.
26. C. S. Lewis, "Reply to Professor Haldane," in *Of Other Worlds: Essays and Stories*, ed. Walter Hooper (New York: Harcourt Brace Jovanovich, 1966), 78. Prior to his conversion to Christianity, Lewis wrote in his diary: "I began to read Haldane's *Daedalus*—a diabolical little book, bloodless tho' stained with blood." (entry for February 20, 1924). See also Lewis, *All My Road Before Me: The Diary of C. S. Lewis 1922–1927*, ed. Walter Hooper (London: Harcourt Brace Jovanovich, 1991), 287.
27. Ibid., 80.
28. Ibid., 80–81.
29. Ibid., 82.
30. Ibid., 82.
31. Ibid., 83.
32. Ibid., 83–84.
33. Ibid., 84.
34. C. S. Lewis, "The Funeral of a Great Myth," in *Christian Reflections* (Grand Rapids, MI: Eerdmans, 1994), 82–93.
35. See also: Mary Midgley, *Evolution as a Religion: Strange Hopes and Stranger Fears* (London: Routledge, 1985). Midgley refers to evolution as "the creation myth of our time" and uses the phrase "the escalator myth" when describing continued human evolutionary development.
36. See Teilhard de Chardin, *The Phenomenon of Man* (San Francisco, CA: Harper Perennial Modern Classics, 2008). Wells also advocated the idea of collective intellect that would achieve conscious self-awareness: "[W]e want a widespread world intelligence conscious of itself. To work out a way to that world brain organization is therefore our primary need in this age of imperative construction. *The World Brain* (Garden City, NY: Doubleday, Doran & Co, 1938), xvi. The idea may have originated with Vladimir Vernadsky (1863–1945) and was adopted by the movement known as Russian Cosmism, an important precursor to Transhumanism. On Cosmism, see: Michael Hagemeister, "Russian Cosmism in the 1920s and Today," in *The Occult in Russian and Soviet Culture*, Bernice G. Rosenthal, ed. (Ithaca, NY: Cornell University Press, 1997), 185–202.
37. Bertrand Russell, *Has Man a Future?* (Baltimore, MD: Penguin, 1961), 13.
38. Nick Bostrom, "Transhumanist Values," April 18, 2001, accessed May 12, 2012, http://www.nickbostrom.com/tra/values.html.
39. See, for example: Allen E. Buchanan, *Beyond Humanity? The Ethics of Biomedical Enhancement* (Oxford: Oxford University Press, 2011); *Human Enhancement*, ed. Julian Savulescu and Nick Bostrom (Oxford: Oxford University Press, 2011).
40. John Stewart, "The Potential of Evolution," *What is Enlightenment?* (February/March 2007), accessed May 23, 2012, . http://www.enlightennext.org/magazine/j35/stewart.asp.

41. Lincoln Cannon, "Trust in Posthumanity and the New God Argument," *Transhumanism and Spirituality Conference* (Salt Lake City, October 1, 2010), accessed May 23, 2012, http://lincoln.metacannon.net/2010/10/transcript-of-presentation-at.html. Many intellectually inclined younger Mormons embrace Transhumanism and find it to comport with their theology. The largest Transhumanist chapter is the Mormon Transhumanist Association in Salt Lake City.
42. Hans Moravec, *Mind Children: The Future of Robot and Human Intelligence* (Cambridge, MA: Harvard University Press, 1988).
43. James Hughes, *Citizen Cyborg: Why Democratic Societies Must Respond to the Redesigned Human of the Future* (Cambridge, MA: Westview Press, 2004), xv.
44. Bostrom, "Transhumanist Values." See also Nick Bostrom, "Human Genetic Enhancements: A Transhumanist Perspective," *The Journal of Value Inquiry* 37 (2003): 493–506, accessed May 26, 2012, http://www.nickbostrom.com/ethics/genetic.html.
45. Hughes, *Citizen Cyborg*, xii.
46. Hugo de Garis, "Globa: Accelerating technologies will create a global state by 2050," January 19, 2011, accessed January 20, 2011, http://www.kurzweilai.net/globa-global-state-by-2050.
47. Hughes, *Citizen Cyborg*, xviii.
48. Bostrom, "Genetic Enhancements," 494–495.
49. Ronald Bailey, "The Democratic Threat to Transhumanism," paper read at "Humanity+ Summit: Rise of the Citizen Scientist" (June 10–12, 2010, Harvard University). For a fuller version of his case, see "Transhumanism and the Limits of Democracy," *Reason.com*, April 28, 2009, accessed May 23, 2012, http://reason.com/archives/2009/04/28/transhumanism-and-the-limits-o/singlepage. Bailey is a writer for *Reason* magazine and author of *Liberation Biology: The Moral and Scientific Case for the Biotech Revolution* (New York: Prometheus, 2005).
50. See Ray Kurzweil, *The Singularity is Near: When Humans Transcend Biology* (New York: Penguin, 2006). Also, *Transcendent Man*, directed by Barry Ptolemy (2011).
51. *Time* magazine interview, Steve Koepp, "10 Questions for Ray Kurzweil," accessed May 12, 2012, http://www.youtube.com/watch?v=0BUYbEgOZt4.
52. John Harris, *Enhancing Evolution: The Ethical Case for Making Better People* (Princeton, NJ: Princeton University Press, 2009), 11. Emphasis in original.
53. Hugo de Garis, "Think About It." (video), accessed December 12, 2011, http://www.youtube.com/watch?v=2A_5-Van9m4.
54. Freeman Dyson, *Imagined Worlds* (Cambridge: Harvard University Press, 1998), 157.
55. Ibid., 156.
56. Hughes, *Citizen Cyborg*, 8.
57. Hava Tirosh-Samuelson, "Engaging Transhumanism," in *Transhumanism and Its Critics*, ed. Gregory R. Hansell and William Grassie (Philadelphia, PA: Metanexus Institute, 2011), 19–20. Other criticisms of human enhancement include Michael Sandel, *The Case against Perfection: Ethics in the Age of Genetic Engineering* (Cambridge, MA: Harvard University Press, 2009); Nicholas Agar, *Humanity's End: Why We Should Reject Radical Enhancement* (Cambridge, MA: MIT Press, 2010); Michael Hauskeller,

"Reinventing Cockaigne: Utopian Themes in Transhumanist Thought," *The Hastings Center Report* 42, no. 2 (April 2012): 39–47.

58. Francis Fukuyama, *Our Posthuman Future: Consequences of the Biotechnology Revolution* (New York: Picador, 2002), 7.
59. Ibid., 7.
60. Lewis, *Abolition of Man*, 49.
61. Ibid., 49–50.
62. Ibid., 48.

11

THE EDUCATION OF MARK STUDDOCK

HOW A SOCIOLOGIST LEARNS THE LESSONS OF THE ABOLITION OF MAN

Cameron Wybrow

IN THE INTRODUCTION TO HIS NOVEL *That Hideous Strength*, C. S. Lewis wrote: "This is a 'tall story' about devilry, though it has behind it a serious 'point' which I have tried to make in my *Abolition of Man*."[1] At the time of Lewis's writing of this remark (Christmas Eve, 1943), the non-fictional *Abolition of Man* had not been out long. Indeed, it had appeared only in late 1943, although the lectures on which the book was based had been delivered in February of that year. It seems that Lewis was working on at least parts of the two works at the same time (late 1942 to early 1943). The overlap in themes is plainly evident to readers of the two works, so there is no reason to doubt Lewis's explanation of the close relationship between them.

To bring out the many connections in thought between *That Hideous Strength* and *The Abolition of Man* would require several essays. In this chapter, I have tried to capture only one set of connections, those which are revealed through the intellectual and spiritual struggle of one of the protagonists of *That Hideous Strength*, sociologist Mark Studdock.

To many readers of the novel, Studdock will not seem like the obvious choice. After all, the most striking philosophical ruminations in *That Hideous Strength*—with phrasing that in some cases seems taken nearly *verbatim* from passages in *The Abolition of Man*—come from the mouths of villainous characters such as Feverstone, Filostrato, and Frost. Yet Studdock's character takes us closer to the educational purpose of *The Abolition of Man* than do any of the purely villainous characters, for, while the villainous characters represent the dehumanizing apotheosis of modern technological society, the character of Studdock reveals the banal yet malignant intellectual and moral roots which produce such a society.

In Studdock we discern how the way in which the members of a society are educated will determine how the society will turn out. More specifically, in Studdock we see many of the opinions and attitudes that motivate the authors of "The Green Book," the upper-year high school English text discussed at length in *The Abolition of Man*. As Studdock's career ambitions move him deeper and deeper into the intrigues of the N.I.C.E., we see him slowly accepting more and more of the implications of the philosophy advocated in *The Green Book*; and in his final rejection of the project of the N.I.C.E., we see his repudiation of that philosophy. In this sense, the story of *That Hideous Strength*, while being an epic treatment of the problem of modern man, is also the story of the failure of modern education, as illustrated by the case of Mark Studdock.

The Main Argument of *The Abolition of Man*

In this essay I venture to presume on the reader's part a familiarity with the plot of *That Hideous Strength*. I therefore do not summarize the novel, but refer to it as if the reader knows the characters and can easily recall the main events in the story. However, I do not presume the same level of familiarity with the contents or argument of *The Abolition of Man*, and, as my purpose is to elucidate the connection between the two without constantly interrupting the flow of my discussion of *That Hideous Strength* with blocks of necessary information from *The Abolition of Man*, I have judged it helpful to the reader to first set forth a brief summary of the argument of *The Abolition of Man*. Those who are already very familiar with *The Abolition of Man* might choose to bypass this brief summary; on the other hand, those who have never read it, or who have not read it in a long time, or have read it but found it difficult, may find that the summary is exactly what they need to enable them to understand the argument of the rest of this essay.

The full title of Lewis's book is *The Abolition of Man; or, Reflections on education with special reference to the teaching of English in the upper forms of schools*.[2] From the second part of this title, one might infer that the book would be mainly of pedagogical interest, an essay on the goals and methods of English instruction in the senior grades ("upper forms") of the British high schools. But the two titles are juxtaposed for a very good reason: Lewis sees, in the defects of the teaching of English literature in the British schools, signs of the decay of long-standing affirmations of the civilization of the West. His critique of English pedagogy will thus pass into a critique of the philosophy which dominates modern Western culture.

The work opens with a discussion of "The Green Book," Lewis's pseudonym for *The Control of Language* (1939) by Alec King and Martin Ketley, a textbook intended for use in the senior grades of the British schools.[3] Commenting that he did "not want to pillory two modest practising school-masters who were doing the best they knew,"[4] Lewis suppresses the authors' real names in favor of the pseudonyms "Gaius" and "Titius."

Lewis zeroes in on the response by Gaius and Titius to a famous story about Coleridge at a waterfall, in which Coleridge endorses the judgment that the waterfall is "sublime":

> Gaius and Titius comment as follows: "When the man said *That is sublime*, he appeared to be making a remark about the waterfall... Actually... he was not making a remark about the waterfall, but a remark about his own feelings. What he was saying was really *I have feelings associated in my mind with the word 'Sublime,'* or shortly, *I have sublime feelings*... This confusion is continually present in language as we use it. We appear to be saying something very important about something: and actually we are only saying something about our own feelings."[5]

Lewis's critique of this statement occupies the rest of the first chapter. Early on in the critique, Lewis writes: "The schoolboy[6] who reads this passage in *The Green Book* will believe two propositions: firstly, that all sentences containing a predicate of value are statements about the emotional state of the speaker; and secondly, that all such statements are unimportant."[7] Against these propositions (and related propositions offered by Gaius and Titius), Lewis urges: "Until quite modern times all teachers and even all men believed the universe to be such that certain emotional reactions on our part could be either congruous or incongruous to it—believed, in fact, that objects did not merely receive, but could *merit*, our approval or disapproval, our reverence or our contempt."[8]

Augustine, Lewis tells us, "defines virtue as *ordo amoris*, the ordinate condition of the affections in which every object is accorded that… love which is appropriate to it."[9] The same notion, he argues, is found in Shelley, Traherne, Aristotle, Plato, early Hinduism, and ancient Chinese traditions.[10] Summarizing this universal tradition, Lewis writes: "This conception in all its forms, Platonic, Aristotelian, Stoic, Christian, and Oriental alike, I shall henceforth refer to for brevity simply as 'the *Tao*'… It is the doctrine of objective value, the belief that certain attitudes are really true, and others really false, to the kind of thing the universe is and the kind of things we are."[11]

There is thus a conflict between the *Tao* and the view advocated by *The Green Book*. This conflict necessarily produces two different views of education:

> Hence the educational problem is wholly different according as you stand within or without the *Tao*. For those within, the task is to train in the pupil those responses which are in themselves appropriate, whether anyone is making them or not, and in making which the very nature of man consists. Those without, if they are logical, must regard all sentiments as equally non-rational, as mere mists between us and the real objects. As a result, they must either decide to remove all sentiments, as far as possible, from the pupil's mind; or else encourage some sentiments for reasons that have nothing to do with their intrinsic "justness" or "ordinacy."[12]

Developing the last point, Lewis argues that for teachers outside of the *Tao*, education can only be: (1) the rigorous "debunking" of all noble sentiments, with the consequent production of a moral vacuum; (2) propaganda, consisting of pseudo-moral principles which, though groundless, are deemed desirable for getting people to act in certain ways.[13]

Lewis grants that the authors of *The Green Book* do not endorse propaganda. This leaves them with the first alternative, i.e., to debunk every noble sentiment in sight, and to presume that there is some biological or other natural basis for morality that does not require noble sentiments.[14] However, in the rest of the chapter, Lewis goes on to show the hopelessness of such an assumption. He argues:

> Let us suppose for a moment that the harder virtues could really be theoretically justified with no appeal to objective value. It still remains true that no justification of virtue will enable a man to be virtuous. Without the aid of trained emotions the intellect is powerless against the animal organism… In battle it is not syllogisms that will keep the reluctant nerves and muscles to their post in the third hour of the bombardment. The crudest sentimentalism (such as Gaius and Titius would wince at) about a flag or a country or a regiment will be of more use.[15]

This insight brings Lewis to the argumentative climax of the chapter:

> We were told it all long ago by Plato. As the king governs by his executive, so Reason in man must rule the mere appetites by means of the "spirited element." The head rules the belly through the chest—the seat, as Alanus tells us, of Magnanimity, of emotions organized by trained habit into stable sentiments. The Chest-Magnanimity-Sentiment—these are the indispensable liaison officers between cerebral man and visceral man. It may even be said that it is by this middle element that man is man; for by his intellect he is mere spirit and by his appetite mere animal.[16]

In the light of his Platonic understanding, Lewis summarizes his criticism of Gaius and Titius thus: "The operation of *The Green Book* and its kind is to produce what may be called Men without Chests." But without "chests,"[17] i.e., without the middle part of the soul that Plato called *thumos*, neither creative energy nor self-sacrifice, neither enterprise nor virtue, is possible.[18] Thus, in seeking to abolish *thumos*, the authors of *The Green Book* unwittingly seek to abolish man.

In the second chapter of *The Abolition of Man*, Lewis notes that, despite the ferocity with which the authors of *The Greek Book* debunk the basis of all judgments of value, they in fact are driven by such judgments themselves: "Gaius and Titius will be found to hold, with complete uncritical dogmatism, the whole system of values which happened to be in vogue among moderately educated young men of the professional classes during the period between the two wars."[19]

Lewis gives something of the content of that "system of values" in his note to the comment: Gaius and Titius disapprove of terms like "brave" and "gentleman," and of the feeling of patriotism. But they approve of peace, cleanliness, democracy, and education.[20] The suburban, middle-class values of comfort and security, rather than either heroic or aristocratic values, are what motivate them.

Having noted that Gaius and Titius are not, for all their debunking, "value-free," Lewis wonders why it is that they think that *their* moral judgments, as opposed to those of more old-fashioned people, should be invulnerable to debunking. He suggests that they imagine that they provide a more "realistic" foundation for morality than the traditional, sentimental basis.[21] However, Lewis is also determined to test the foundations of this allegedly more realistic morality, and by the end of chapter 2, he has shown that all the alleged non-sentimental arguments (e.g., utilitarian arguments) for Gaius and Titius's particular set of moral values can be "seen through" or "debunked" just as surely as can the noble sentiments connected with the *Tao*.[22]

This conclusion sets up Lewis's third and final chapter. Having shown that the "debunking" spirit of *The Green Book* is inconsistent with the maintenance of any firmly held moral system or stable society, Lewis considers what might happen if teachers like Gaius and Titius were to abandon their inner inconsistency and reject the concept of value altogether.

Lewis envisions a future society in which "man's conquest of nature" is complete. Human beings have mastered external nature and can use it to produce food, shelter, medicine, and infinite creature comforts and entertainments at will. The survival and physical contentment of the race is thus guaranteed. But mastery over external nature will eventually involve mastery over biological and psychological nature, as the tools of science become more probing and powerful; thus, mankind will be able to remake, not only the external environment, but even itself. The human race will be able, for the first time, to define itself, to create its own future reality. This creates the paradoxical situation in which man is at once master and slave: master over the externalities of nature—heat, cold, hunger, sickness, etc.—but enslaved by the science which will control his own genetic constitution and his own emotional makeup through high-tech biological and psychological manipulation. And who will be in charge of that scientific manipulation?

Lewis foresees a class of men called "the Conditioners." The Conditioners, more consistent than Gaius and Titius, have "seen through" all attempts to ground behavior in any ultimate truth about the way things are, and consequently have rejected the authority of the *Tao* or any fragments of it (such as the liberal, humanist fragments which still guide Gaius and Titius). The *Tao*-less Conditioners are charged with deciding what man is to be. They will choose the physical talents of future generations of human beings, and they will choose the set of emotions and valuations to implant in those future generations, in order to get them to behave in a desired way. But Lewis asks the question: What will guide the Conditioners in their choice of the "values" which future men will have?

These Conditioners may at first, out of habit, inconsistently rule under the sway of the old-fashioned morality, i.e., under the impression that as rulers they have some obligation of stewardship, some duty to promote the good of those ruled. Thus, there might be a period of the government of the many, by the few, for the many. But this state of affairs, Lewis points out, cannot last.[23] At some point the Conditioners will fully realize that their sense of obligation was nothing more than the result of the social conditioning of previous societies, and will cast it off; after that, the only thing governing their behavior will be their current whims. They will make the kind of human beings that they please, and condition them with drugs and education and propaganda as they please. With the last traditional moral restraint gone, the mastery of the human race over nature will turn into the mastery of the few over the many. It will be government of the many, by the few, for the few.

What will control how the few will act? Will they act, as in *Brave New World*, with the goal of giving the many a life of comfortable convenience, spiced with erotic and drug-induced pleasures? Or will they act, as in *1984*, with the goal of enslaving and crushing the human spirit? There is no way of predicting, since the Conditioners will act on the basis of their strongest current impulses. Those impulses are as likely to be bad as to be good. And where will those impulses come from? They cannot come from the *Tao*, which the Conditioners have rejected as an artifice corresponding to no ultimate reality. They will then come from chance factors that affect the rulers, e.g., "heredity, digestion, the weather, and the association of ideas."[24] That is, they will come from unregulated Nature. The rulers, having repudiated the *Tao*, will let the previously subdued Nature back in, through the doorway of their own unguided wills. And since their wills are the ruling power over their civilization, their civilization, and all the human beings within it, will be ruled by the random impulses of nature—a massive irony for a civilization which started out in hopes of liberating itself by the conquest of nature.[25] And this civilization, if technologically competent, will never be overthrown: "Nature will be troubled no more by the restive species that rose in revolt against her so many millions of years ago, will be vexed no longer by its chatter of truth and mercy and beauty and happiness... and if the eugenics are efficient enough there will be no second revolt, but all snug beneath the Conditioners, and the Conditioners beneath her, till the moon falls or the sun grows cold."[26]

In such a situation, man himself will have been abolished. The Conditioners, Lewis says, will be neither good men nor bad men; they will not be men (in the true sense) at all. To be "men" in the old sense, they would have to be within the *Tao* (which determines what it is to be human, and what it is to deviate from the human), and thus to be capable of feeling the shame which comes with such deviation. But they are not within the *Tao*. They lack the middle part of the soul by which impulses are controlled in the name of something higher than impulse; they are, in Lewis's image, "men without chests," and hence, outside of humanity altogether.[27]

And what of the conditioned? They will not live under the *Tao*, but only under whatever substitute moral principles the Conditioners have instilled in them. Still, to the extent that such moral principles are internally consistent, might they not be said to participate in some dim way in humanity? They will have at least an "artificial conscience," unlike their rulers and masters.[28] Yet Lewis's judgment upon them is clear: "They are not men at all: they are artefacts."[29] Thus, the civilization which shall have completely conquered nature (and in turn become completely enslaved by it), consists of rulers who, being beyond good and evil, are above (or at least outside) humanity, and subjects who are beneath it.[30]

Such, Lewis teaches, is the fate of the human race if it consistently abandons the authority of the *Tao*: a mastery over nature which puts man under the heel of nature to a greater extent than ever before; a global tyranny of the conditioners over the conditioned, and the abolition of man himself. And since the abandonment of the *Tao* is encouraged and justified by the kind of arguments against noble sentiments which the authors of *The Green Book* would make a standard part of public education, the abolition of man will be the ultimate outcome if the philosophy and pedagogy of *The Green Book* is allowed to stand.

The Education of Mark Studdock in *That Hideous Strength*

MARK AND JANE STUDDOCK ARE in a sense the protagonists of *That Hideous Strength*. They are an unhappily married young couple, who, in the course of the story, are forced to come face to face with the faults each of them has which have caused their marriage to be so unhappy. The events of the story separate them for most of the book, and the sort of challenges that they face, while in some ways parallel, are also specific to each character's weaknesses, and the particular set of weaknesses possessed by Mark are the ones which concern us here, as they connect much more directly with the themes of *The Abolition of Man*. Thus, without implying that Jane is less important in the story than Mark, we will focus almost exclusively on Mark.

Mark is a young sociologist. His exact age is not given, but as Jane is said to be twenty-three, it is a reasonable guess, given the age difference between husband and wife typical of his class and era, that he is about twenty-eight. In confirmation of this estimate, we know that he has been at Bracton College for at least a few years, long enough to be well-acquainted with its internal politics; and if we assume that he started teaching there at about age twenty-three,[31] an age in the late twenties seems quite likely.

Mark's exact area of research in sociology is not given, in part, no doubt, because it is not relevant to the story, but perhaps in part because Lewis wants us to see Mark's social thought as broad and diffuse rather than focused and incisive. If Mark were a precise sort of sociologist, e.g., the type who does detailed mathematical analyses of particular social phenomena, his thinking would have a disciplined and scholarly character which the narrator tells us it did not have. We know that he lacks both scientific and Classical training,[32] and that he was a "glib examinee in subjects that require no exact knowledge," who "had always done well on Essays and General Papers."[33] We are thus not to imagine Mark as a Max Weber or an Emile Durkheim, a sociologist steeped in the rigors of Continental scientific and humanistic learning. He is more like the "socially relevant" sociologist of today, an all-purpose social analyst, inclined to make large generalizations about the nature of man and society based on the currently dominant set of social science concepts (e.g., "class"), but not terribly critical of the theoretical foundations of his subject.

Like many sociologists then and now, Mark has an interest in his subject not primarily out of the sheer love of knowing—which was the motive of Greek *theoria*—but because he sees it as preparing the way for a more "scientific" society: a more organized, methodical form of human living which, being more efficient, will produce wealth, health, inventions, progress, and happiness in a systematic way not possible in previous societies. Thus, he sees his studies as aimed at the social good, in a way that (to his mind) dusty old Classical studies like Greek and Latin are not. (Mark's school Latin is represented in the story as rudimentary and practically all forgotten.)

In the course of the story, Mark is lured away from his teaching position at Bracton College to become part of the massive research effort known as the N.I.C.E. (National Institute for Co-ordinated Experiments). The N.I.C.E. is a "think tank," as we should now say, but a think tank with an extreme technological bent. It desires to see scientific knowledge applied not only to external things (better fertilizers and radios and the like), but to the control and improvement of human behavior. It is thus heavily into intrusive biological and psychological research (including experimentation upon criminals and mental patients), research which takes place largely behind closed doors and out of the public eye. It champions the new (and purportedly more humane) idea that the purpose of a legal system is not for the meting out of just punishments, but for the rehabilitation of the offender; thus, it arranges for the State to ship a good number of criminals to its laboratories for social reconditioning. Its long-term goal is to have *all* criminals taken out of the hands of traditional authorities (police, courts, and jails) and turned over to N.I.C.E. researchers for inner reshaping. But it is not merely criminals that it wants to reshape; it hopes in the long run to re-socialize all of England, by applying "scientific" measures of control and education, ultimately even prenatal conditioning, to bring into being a new and improved form of humanity.

At least, these are its overt goals. As the story proceeds, we learn that beneath this scientific humanism, the demon-controlled leadership of the N.I.C.E. intends eventually to do away with humanity altogether, and replace it with a small number of highly evolved beings who have none of the softness or sentimental characteristics we associate with human beings, because they have evolved into nearly pure minds, free of bodily constraints. But only the inner circle of N.I.C.E. people knows of this goal. The majority of its functionaries and researchers are kept in the dark, and led to believe that they are collaborating in a grand secular humanist project, an extended Baconian mission to relieve the human estate through better food and medicine and to improve both individuals and societies by remaking them through modern technology. When Mark is first brought into the N.I.C.E. by Lord Feverstone, this is his own understanding of the project. As the story progresses, he is slowly brought deeper and deeper into the inner ranks of the organization, and eventually the supernatural motivation behind its workings is made known to him.

Mark is at first wanted by the organization because he is a good writer. He has a gift for turning sociological insights into persuasive popular prose. His mission is to propagandize the public to accept the idea of a more and more scientific society, with the state (or quasi-state institutions like the N.I.C.E.) having more and more power in traditional realms such as law, child-rearing, economics, and education. His job is to convince the public that increasing central control over such things, far from being a threat to human freedom, will increase human freedom and happiness by providing more wealth, more health, greater longevity, more technical gadgets to make life more comfortable, etc. This talent of his for propaganda is used by the Institute, and gradually he is asked to write things that are more and more dishonest, knowingly concealing from the public the true extent of the power over individual lives that the N.I.C.E. seeks. Mark is praised by the inner circle of the N.I.C.E. for his artfulness in the expression of deceptive half-truths, and through the pride generated by this praise, Mark is corrupted.

Mark was in fact picked out by the N.I.C.E. very early on—they had even influenced his earlier hiring at Bracton College—precisely because he was felt to be a gifted writer whose morality was somewhat malleable. But this must not be misunderstood. Mark was not dishonest about matters of money or sexual fidelity. He thought of himself as, and in many ways was, a conventionally moral person. But he was young, academically vain, eager to please people, and eager to be thought well of, and the inner circle of the N.I.C.E. was looking for just such a person. A person incorruptible by crude means may be quite corruptible by more insidious ones—those which play upon pride and ambition.

Within Mark's discipline of sociology, one of the ways to "move up" in the esteem of one's colleagues is to show how one can "see through" all the institutions of society, how they serve the selfish purposes of certain groups or classes, rather than the good of the whole. Thus, Mark has contempt for the class of *rentiers*, the small investors, whom he calls a "bad element";[34] he has learned from his sociology to think of them as a selfish class, which ought to be swept away by a scientific reorganization of society. He is (in line with much twentieth-century sociological and political science opinion) less harsh toward the peasant and the working classes, but he soon learns that the N.I.C.E. regards them, too, as a retrograde influence that stands in the way of progress, and that they, too, must be eliminated or reeducated.[35] He is thus to lend his pen to the propaganda which will eventually subject these people to the schemes of the N.I.C.E. for a new, scientific, planned England.

Mark's sociological habit of "seeing through" classes and their weaknesses clashes with a certain instinctive knowledge which he possesses (though in much smaller quantity than does his wife Jane, or any well-balanced, decent Englishman). When he visits a village full of bourgeois and working-class people, he finds that they are in many ways likeable; his lived experience does not fit in with his political/sociological "deconstruction" which "proves" these people to be a dead weight upon society.

But over his weeks at the Institute, he slowly learns to suppress all such instinctive and empirical knowledge in favour of the abstract conceptions (class, etc.) which he is so clever in expounding. He must be against what the Institute is against, and for what it is for, if he hopes to rise high in its ranks and be influential in it.

Mark has other personality traits which are of use to the N.I.C.E. For example, he is "not as a rule very sensitive to beauty." This is important, as the N.I.C.E. will be doing many things that are not beautiful, from cutting up living animals (Mark has no objections to vivisection[36]), through tearing down the ancient Bragdon Wood at Bracton College, to destroying the quaint village of Cure Hardy in order to reroute a river for the N.I.C.E.'s purposes. A N.I.C.E. colleague of Mark's, Cosser, thinks that the cawing of the rooks at Cure Hardy, on the fine sunny day that they visit the village, is a "bloody awful noise."[37] The N.I.C.E. has no love of nature as such; it wants to master nature, beat it into submission for its own ends. Therefore, preserving the beauty of nature—or of social institutions—is of no concern to it.

But it is not merely the political program of the N.I.C.E. which requires Mark to stamp out of his soul any fleeting concerns he has for beauty. The N.I.C.E., which is secretly run by evil *eldila*, who are in fact demonic beings, aims at more than political and social control of England; indeed, political and social power is only a subsidiary end. It ultimately desires the conversion of the souls of the elite personnel of the N.I.C.E. to demonic ends. As Mark is brought into the inner circle, he is eventually invited into the deepest secrets of the organization, and the initiation will require him to perform a series of actions involving the embrace of filth and ugliness and the rejection of wholesomeness and beauty. He is of course prepared for this more than most people by his education, which has been "Modern"; whereas a Classically educated person would be moved by beauty and unwilling to damage it, someone educated in the modern view—that all positive reactions to beauty are mere statements of private emotion having no truth-content, would be much more willing to desecrate the beautiful in order to work his way into the center of an organization. Thus, the extended discussion of the beautiful and the sublime with which *The Abolition of Man* opens is central to Mark's temptation. It is precisely because he has a modern education, of the sort pushed in *The Green Book*, that he has a contempt for beauty, and therefore might be induced to repudiate it. And of course, as the demons know, the love of the beautiful and the love of virtue are connected; to weaken the sense of the beautiful is to weaken the sense of the good overall, and hence to weaken virtue. And to become a full initiate of the N.I.C.E., Mark must transcend virtue altogether.

Thus, in Mark, Lewis paints for us a portrait of the "modern" educated man as set forth in *The Green Book*—largely insensitive to beauty, and tending to "see through" all alleged virtues (whether of individuals or social classes or societies) and dismiss them as emotional rubbish which must be swept away in the name of rationality and progress. In the light of *That Hideous Strength*, we can see that the authors of *The Green Book* are offering a recipe for producing more Mark Studdocks, and not just in the Sociology departments but in the English and other humanities departments, and in fact in all walks of life, since all, or certainly many, will pass through senior English in the schools. Their principles would generate a group of young intellectuals who mock virtue and beauty and cannot exemplify them. Such people would become, as it were, the willing disciples of demons, since the teachings they would follow would be, if not literally demonic in origin, certainly demonic in effect.

Of course, we must not suppose that Mark, despite his lack of understanding of beauty and virtue, is himself a vicious man or a man without some set of values.[38] Mark has a set of values. He believes in equality, for example, and is surprised when Professor Filostrato and the Rev. Straik speak against it in the name of the supremacy of a single superman.[39] Like all good democrats and socialists of the modern world, he also believes in the use of science to increase the wealth and happiness of the masses,[40] and in the goal of universal education.[41] Finally, Mark believes above all in the survival of the human race.[42] But in believing these things Mark is, as Lewis says of Gaius and Titius, better than his principles.⁴³ For Mark has no justification for his beliefs in progress, universal education, universal prosperity, equality, democracy, or even racial survival.⁴⁴ The sentiments underlying these beliefs are just as emotional and ungrounded as the sentiments for beauty and for kindness to animals which Mark rejects. He is an inconsistent modern man; a modern man who "sees through" certain older conceptions of beauty and truth and social good, but who does not "see through" his own views. We are thus reminded of what Lewis said in *The Abolition of Man*: "Gaius and Titius… hold… with complete uncritical dogmatism, the whole system of values which happened to be in vogue among moderately educated young men of the professional classes during the period between the two wars." It is only Mark's exposure to the inner circle of N.I.C.E.—first Feverstone, then Filostrato and Straik, and finally Frost—that eventually teaches him that even his wised-up, liberal, progressive values have no more truth than the dusty old Classical and Christian values which he has rejected.

It is in the conversations between Mark and Frost that Mark gains final clarity about the status of his own cynical debunking of tradition, and where it leads. For in Frost we see the radical completion of modern thought, as set forth by Lewis in *The Abolition of Man*. At the end of chapter 2 of that work, Lewis had established that the *half-hearted* rejection of standards of value—the rejection of parts of the *Tao*, while retaining such parts of it as were necessary to sustain liberalism, progress, democracy, and so on—was logically inconsistent and could not stand. But in the same passage, he granted that a *full* rejection of the concept of objective value—the abandonment of any attempt to seek out, and adhere to, the good, the true, and the beautiful—could be a consistent position.⁴⁵ In *That Hideous Strength*, while the adoption of that position is characteristic of most of the inner circle of the N.I.C.E., its fullest incarnation is found in the character of Frost.

Frost is wonderfully characterized in the book. His very name suggests an icy aloofness from others, and indeed, he exemplifies that throughout the story. His appearance—pointed beard and *pince-nez*—

evokes the stereotype of the cool, clinical psychoanalyst who regards his patients with supreme objectivity. His conversation with all members of the N.I.C.E., even his superior Wither, is laconic, precise, and without wasted words, always driving to the point without concern for diplomacy or the esteem of others. When he converses with Mark in the prisoner's cell, where Mark is being held under threat of execution, he reveals the deepest secrets of the N.I.C.E.—the organization to which he has given his soul and into which he wants Mark to be fully initiated—and there is not a hint of emotion: "Frost remained, throughout the whole conversation, standing perfectly still with his arms hanging down straight at his sides. But for the periodic upward tilt of his head and flash of his teeth at the end of a sentence, he used no gestures."[46] In comparison with Frost, Mr. Spock from Star Trek is an ocean of seething emotions.

Frost believes in "objectivity." For him, objectivity starts with the recognition that resentment and fear—and love and all other emotional responses—are merely chemical phenomena. A rational person does not let himself be ruled by chemical phenomena. He rises above them, not by suppressing them by an act of will, but by realizing their non-objective character. Once their purely subjective nature is fully grasped, they no longer have power over one; one is liberated. This means, of course, that to be truly free, one must leave behind not only all the negative emotions—hatred and jealousy and fear—but all the positive emotions as well, and the whole world that goes with them: the world of virtue and beauty and morality. One must rise to a plane that is beyond good and evil as human beings normally understand those terms. Mark's final initiation involves the acquisition—through a rigorous training program designed to destroy all emotional attachment—of this form of objectivity. If he passes his initiation, he will be, like Frost, free of all traditional human motivations. Mark is thus invited to examine the vision of existence painted by Lewis in the last part of *The Abolition of Man*. His choice whether or not to continue with his initiation will depend on how he assesses that vision.

It would be misleading to overlook the fact that Mark's final decision is influenced by things other than the intrinsic merits of Frost's philosophy. Mark has, by the time that Frost first offers him initiation, become disillusioned in many ways regarding the N.I.C.E. He has learned that Feverstone brought him into the Institute not out of old college friendship but because his propaganda skills would be useful. He has learned that the N.I.C.E. has murdered his colleague Hingest, and planted his wallet near Hingest's body so as to make him look guilty of the murder. He has learned that Miss Hardcastle, head of the N.I.C.E. police, burned his wife Jane with cigarettes in an interrogation. He has been manipulated emotionally all along by Wither, the Deputy Director. He has come to see that, even as he was apparently being brought more and more into the inner circle of the organization, he was more and more being used and enslaved by it. And he now knows that the Institution wants to gain possession of Jane, for dark reasons that he cannot imagine (he is not aware that they want her for her gift of clairvoyance), and he is not yet so far gone in evil that he wishes to offer up his wife as a sacrifice to such people. So he is already well on his guard by the time Frost comes to him with an offer of salvation both physical—release from jail and from the death sentence for the alleged murder of Hingest—and spiritual—admission into the innermost ring of N.I.C.E., the small group that communicates directly with the bent *eldila*, the demons who are running the whole operation behind the scenes.

Nonetheless, Mark does, in the course of his conversations with Frost, and in the course of the initiatory acts which he performs (in order to fool Frost into thinking that wishes to complete the process), reason critically about the theoretical basis of Frost's position. He probes the arguments of Frost with a clarity that exceeds the blurry fuzziness he practiced as a sociologist, and he probes the philosophical position that he himself has held for years and which has led him to where he now is. Thus, while the manipulations and humiliations he has suffered, and the repulsiveness of the personnel of N.I.C.E., are major factors in his final decision, the decision is clinched by argument—an argument which in fact is identical with the argument of the closing chapter of *The Abolition of Man*.

Frost has told Mark shocking things. He has said that war is not, as Mark supposed, wasteful and evil, but in fact a necessary cleansing mechanism to reduce surplus human population, especially the inferior members of the human population—which is why the Macrobes (the evil *eldila*) have planned sixteen major wars for the 20th century.[47] Frost has even denied that the survival of Man as such is important. If Man should pass out of existence, having evolved into something purely mental—all "head" and no "chest," and hence quite beyond what we now call human—that is good, because that is what the evolutionary process ordains.[48] In this context, Frost even refers approvingly to Waddington,[49] a real-life biologist who had written a book about evolutionary ethics—a book whose doctrines Lewis criticized in *The Abolition of Man*.[50] Frost's assertions here are shocking precisely because he has passed beyond the half-hearted moral skepticism advocated by *The Green Book* and achieved the radical amorality of the Conditioners discussed by Lewis in *The Abolition of Man*.

At this point in the dialogue, Mark becomes almost the spokesman for Lewis's own position; he starts to ask questions which strike at the heart of Frost's position. He grants (feigning agreement with Frost) that all traditional ideas of patriotism and humanity are mere product of the animal organism, and irrational; but, he asks, if traditional emotional motives are to be entirely abandoned, what will replace them as the spring of action?[51] Frost's recapitulation and clarification is revealing:

> Motives are not the causes of action but its by-products. You are merely wasting your time by considering them. When you have obtained real objectivity you will recognize, not *some* motives, but *all* motives as merely animal, subjective epiphenomena. You will then have no motives and you will find that you do not need them. Their place will be supplied by something else which you will presently understand better than you do now. So far from being impoverished, your action will become much more efficient.[52]

What does Frost mean by "something else"? It would appear, from the descriptions throughout the book of the behavior of the full initiates Frost and Wither, that he means the direct communication of the will of the Macrobes, the evil *eldila*. The will of the *eldila*, made known through direct verbal communication or through non-verbal promptings, guides the true initiate. Freed of the influence of the organic body, the initiate is moved by pure spirit. Mark has as yet no understanding of this kind of motivation, nor will he, until he has killed all the lower motivations and become a full initiate. But such motivation has problems of its own which Frost does not mention.

Late in the story, after uttering a cold threat to Merlin (whom Frost at that point takes to be a Basque priest), Frost does something very strange: "Suddenly, not as if he wished to, but as if he were a machine that had been worked, Frost kicked him."[53] This action is uncharacteristic of Frost, who, true to his name, is usually "cool" and does not lose his temper. Has Frost, in his impatience with the supposed priest, "fallen off the wagon" and descended into the realm of the emotional? It appears not. Frost acts as if he had been "worked." His body acts as a tool in the hands of another. But what other? Presumably, the Macrobes to whom Frost has given his soul. It is the malice of the Macrobes which, through the body of Frost, kicks Merlin. Frost neither intended nor personally executed the action.

More light is shed on this event by a later scene:

For many years he had theoretically believed that all which appears in the mind as motive or intention is merely a by-product of what the body is doing. But for the last year or so—since he had been initiated—he had begun to taste as fact what he had long held as theory. Increasingly, his actions had been without motive. He did this and that, he said thus and thus, and did not know why. His mind was a mere spectator... Still not asking what he would do or why, Frost went to the garage... He came up with as many petrol tins as he could carry. He piled all the inflammables he could think of together in the Objective Room. Then he locked himself in... Like the clockwork figure he had chosen to be, his stiff body, now terribly cold... poured out the petrol and threw a lighted match into the pile.[54]

Frost ends his life in suicide, and it is a suicide directed by neither his mind nor his will. It is motiveless, as far as he is concerned, yet not ultimately motiveless, for it is imposed upon him from the outside, by the evil *eldila* whom he had faithfully served.[55]

This incident can perhaps best be understood if it is translated out of the demon-filled world of *That Hideous Strength* into the demon-free analysis of *The Abolition of Man*. Lewis tells us that the Conditioners who are beyond the *Tao*, beyond good and evil, and hence beyond all normal human motivation will still be driven by *something*, i.e., by the strongest chance impulse that happens to impinge upon their wills. Just as the initiates in *That Hideous Strength* experience urges to destructive and self-destructive behavior springing from the malignant and ultimately irrational will of the *eldila*, so the Conditioners in *The Abolition of Man* will do whatever they do (to the human beings they rule) out of the disorderly impulses which come to them from blind Nature. If modern society ever reaches the stage of a world ruled by scientific Conditioners, humanity will be as enslaved by irrational caprice as it would be if it were ruled by literal demons.

Thus, in the question about the grounds of action posed by Mark Studdock to Frost, and in Frost's answer (as interpreted in the light of Frost's slavelike and finally suicidal actions), we see the ultimate refutation of Frost's view of "objectivity" and its corollary of motiveless action. Motiveless action, if it could exist, would not be free action, but capricious action. And it would proceed not from human choice but from the random impulses of nature. Far from advocating a higher stage of rational existence than the human, Frost is advocating something lower than the human, a kind of radical un-freedom, a slavery to impulses over which human beings finally have no control, because they have given up the rationality by which they could exercise such control.

There is a moment in the story at which Mark's theoretical awareness becomes complete. Immediately after the statement by Frost to

Mark about the superiority of motiveless action, Mark's thoughts are relayed to us by the narrator:

> The philosophy which Frost was expounding was by no means unfamiliar to him. He recognized it at once as the logical conclusion of thoughts which he had always hitherto accepted and which at this moment he found himself irrevocably rejecting. The knowledge that his own assumptions led to Frost's position, combined with what he saw in Frost's face and what he had experienced in this very cell, effected a complete conversion. All the philosophers and evangelists in the world might not have done the job so neatly.[56]

It is "the knowledge that his own assumptions led to Frost's position" that drives the truth home for Mark. He sees that the intellectual habits of his whole life—of mocking traditional values, of belittling traditional institutions, of ignoring beauty, of regarding nature as purely stuff to be used or an enemy to be conquered, of preferring glib abstractions to the concrete reality of living people, of thinking that a reductionist "science" is the source of all truth and more important than humanity itself—must lead to the insane doctrine of amoral, motiveless action which produces ex-human monstrosities like Frost and Wither, and institutional monstrosities like the N.I.C.E.

As a sociologist, Mark was an educator in the spirit of Gaius and Titius, the authors of *The Green Book*. In his final repudiation of the teaching of Frost, Mark in effect repudiates the position of Gaius and Titius. In that sense, *That Hideous Strength* depicts a sort of wish-fulfillment on the part of Lewis: a Studdockian conversion of the Gaiuses and Titiuses of the world might be effected if such educators would read *The Abolition of Man* and repent of their destructive teaching about noble sentiments and judgments of value.

Concluding Thoughts

WHY IS IT THAT THE Gaiuses and Titiuses of the world do not come to the realization that Mark Studdock arrives at? I think the answer, in part, is that Gaius and Titius, and a thousand educators like them (generally armed, these days, with Ph.D.s in Education), operate in a world of pure theory, a world of academic discourse in which the like-minded preach to each other. Mark Studdock was like that, early on in *That Hideous Strength*. As a fellow of Bracton, he lived in an ivory tower, where the schemes and plots of academic factions were his only reality. But the events in *That Hideous Strength* force Mark to come to grips with reality. They force him to see what effects ideas have when they get out into the real world of political and social and ethical life. The truth of experience enables him to see the falsehood of bad theory. Our modern academics, on the other hand, have undergone no such life-transforming experiences. Living in an almost completely artificial intellectual environment, protected by tenure, and in no way answerable for the real-life effects of their theories, they can offer teachings about sentiments and morals which, however destructive to society in the long term, are unlikely to rebound upon them in the visceral way that Mark's teaching rebounded upon him.

Another part of the answer, I think, is that our theories of education are almost completely dominated nowadays by the social sciences, with very little input from the traditional humanities. And the social sciences are largely oriented to changing the world, in a certain direction which we might politely call "progressive." This deep commitment to social science as a theoretical basis for social change (as natural science is conceived to be a theoretical basis for technological change) makes modern education theorists generally impatient of old-fashioned humanities education. Why waste time in the hopeless, idealistic task of trying to produce better *people* (even if there were an objective definition of "better," which most social scientists deny) when you could be learning the scientific principles that will enable us to control our future social environment, making *society* better even though people remain as intellectually and morally flawed as ever? This, of course, is exactly the opposite of the vision of education held by Lewis. Education must be based on the cultivation of the mind and heart of the individual, and that cultivation must consist of exposing the students to the best that has been thought and felt and known by human beings across the ages.

If there is any hope, then, of preventing what Lewis called "the abolition of man," it will not come from our certified experts on education. They are nearly all unrepentant Mark Studdocks who partake wholly of the very spirit of modernity that Lewis was criticizing. And their insulation within professional cliques of "education theorists" who speak academese rather than English, and who have almost no meaningful contact with the average member of society, guarantees that very few of them will ever have the life-transforming experience of a Mark Studdock. So the necessary counter-teaching must come from elsewhere. It must come from the few remaining real philosophers (most of whom do not teach in philosophy departments), from the few remaining deep theologians (most of whom do not teach in theology departments), from the few remaining inspired poets and dramatists and novelists and other artists, from the few remaining wise historians, and from the (hopefully not so few) remaining sober, sane, well-balanced parents and citizens who continue to know that we are still under the *Tao*. These people will still continue to place humanity above abstractions such as "the conquest of nature," "science," and "progress."

To be sure, such people are in a weaker position to influence the direction of public life than they were in Lewis's time. The institutional supports which still existed, in some measure, in Lewis's day, are now almost entirely gone. The chances that a high school student of today will learn Latin, or read some Herodotus, or will even hear the names of Augustine or Aristotle, are extremely low. The chances that one's local clergyman has a degree in Classical Civilization or Philosophy are similarly low. The Bible and the theological tradition have been steadily pushed out of the public sphere. Teenagers are far more likely to learn "The Five Stages of Grief" than the Seven Deadly Sins or the Four Cardinal Virtues. We are faced with the daunting task of trying to restore the premises of a nearly shattered Western culture.

Nonetheless, despair is never constructive. There are things we can do. We can keep our eyes peeled for the appearance of wise philosophers, wise poets, wise politicians, and wise citizens, and encourage and support them when we identify them. We can financially support the few remaining journals that have contents of moral and intellectual depth. We can expose our friends and families to the best that we know of the ancient religious and philosophical traditions of the West. We can have dinner-table conversations about Socrates or Homer or the Bible instead of about the latest "reality" shows. We can send our children off to a small, intimate liberal arts college to study the Great Books instead of to a mammoth state university to study Abnormal Psychology or Marketing. If we are teachers of history or literature on any level, we can deftly insert material into our courses about historical people and important writings which the prescribed curricula have omitted. And of course, we can encourage those who would benefit from it to read *The Abolition of Man* and *That Hideous Strength*. What was possible for Mark Studdock—the realization that much of contemporary Western civilization is built upon lies and errors regarding what is ultimately true—is still possible for any thoughtful person who reads these books.

Endnotes

1. C. S. Lewis, *That Hideous Strength: A Modern Fairy-Tale for Grown-Ups* (London and Sydney: Pan, 1983), 7. All references in the present essay are to this edition, which contains the unabridged text of the original publication [London: John Lane (The Bodley Head) Ltd, 1945]. The references will use the short form of the title (or, where appropriate, "Ibid."), and will include the chapter and section numbers as designated by Lewis, followed by the page numbers of the Pan edition; e.g., *That Hideous Strength*, ch. 9, sec. 2, 187.

2. C. S. Lewis, *The Abolition of Man; or, Reflections on education with special reference to the teaching of English in the upper forms of schools* (New York: Touchstone, 1996). All references in the present essay are to this edition. The work consists of three unsectioned chapters, an appendix, and endnotes. The references will use the short form of the title (or, where appropriate, "Ibid."), and will include the chapter number, followed by the page numbers of the Touchstone edition; e.g., *The Abolition of Man*, ch. 2, 49. *The Abolition of Man* first appeared in the form of a series of lectures at King's College, Newcastle-upon-Tyne [now part of the University of Durham]. The lectures were delivered in February 1943 and published in book form late in that year.

3. Lewis's copy of this book is now housed at the Wade Center at Wheaton College.

4. *The Abolition of Man*, ch. 1, 17.

5. *The Abolition of Man*, ch. 1, 18.

6. Lewis throughout the book seems to have in mind what the British call "public schools" (which in North America would be called "private schools"); such schools in Britain

in his day were always (or almost always) "boys only" or "girls only"; Lewis has in mind such boys' schools as he himself attended; hence it is only "boys" of which he writes.
7. *The Abolition of Man*, ch. 1, 19.
8. Ibid., ch. 1, 27–28.
9. Ibid., ch. 1, 28–29.
10. Ibid., ch. 1, 28–30.
11. Ibid., ch. 1, 31.
12. Ibid., ch. 1, 32–33.
13. Ibid., ch. 1, 32–34.
14. Ibid., ch. 1, 34–35.
15. Ibid., ch. 1, p. 33.
16. Ibid., ch. 1, 35–36.
17. Ibid., ch. 1, 36.
18. Ibid., ch. 1, 37.
19. Ibid., ch. 2, 42–43.
20. Ibid., n. 1 to ch. 2, 43.
21. Ibid., ch. 2, 43.
22. Ibid., ch. 2, 43–62.
23. Ibid., ch. 3, 72.
24. Ibid., ch. 3, 76.
25. Ibid.
26. Ibid., ch. 3, 77.
27. Ibid., ch. 3, 73–74.
28. Ibid., ch. 3, 75.
29. Ibid., ch. 3, 74.
30. It is hard not to see in Lewis's sketch of a future society the scathing portrait, first sketched by Nietzsche, of a world of "last men" ruled by an elite of "nihilists". Indeed, Lewis's conception of Frost (in *That Hideous Strength*) seems inspired by Nietzsche's portrait of the nihilist.
31. It was not uncommon for British dons of the era to begin teaching with only a Bachelor's degree, with the Master's degree being later granted by a semi-automatic procedure.
32. *That Hideous Strength*, ch. 9, sec. 2, 185.
33. Ibid.
34. Ibid., ch. 4, sec. 6, 85.
35. Ibid.
36. Ibid., ch. 5, sec. 1, 102.
37. Ibid., ch. 4, sec. 6, 87.
38. I hesitate to use the term "values," because as commonly used, it has a Nietzschian flavor, suggesting that "value" is something posited by an act of will of human beings, rather than objectively existing. However, Lewis himself uses the term, so I defer to his usage,

while reminding the reader that "value" in Lewis's language has an objectivity that it does not possess in modern historicist thought.
39. *That Hideous Strength*, ch. 8, sec. 3, 178.
40. Ibid., ch. 12, sec. 4, 258.
41. Ibid., 259.
42. Ibid., ch. 2, sec. 1, 41.
43. *The Abolition of Man*, ch. 1, 34.
44. Note that even Lord Feverstone is skeptical of arguments based on an alleged moral obligation to the future human race; see *That Hideous Strength*, ch. 2, sec. 1, 41–42. Lewis expresses Feverstone's skepticism in an extended argument in *The Abolition of Man*, ch. 2, 51–55.
45. *The Abolition of Man*, ch. 2, 61–62.
46. *That Hideous Strength*, ch. 12, sec. 4, 257.
47. Ibid., ch. 12, sec. 4, 258–259.
48. Ibid., ch. 14, sec.1, 296.
49. Ibid., ch. 14, sec. 1, 295.
50. *The Abolition of Man*, n. 3 to ch. 2, 50. Note that Lewis, observing Waddington's apparent unease with his own conclusions, suggests that Waddington was a better man than his principles seemed to justify; Frost, on the other hand, has adopted Waddington's principles without Waddington's hesitations or equivocations.
51. *That Hideous Strength*, ch. 14, sec. 1, 295.
52. Ibid., ch. 14, sec. 1, 296.
53. Ibid., ch. 15, sec. 3, 332.
54. Ibid., ch. 16, sec. 6, 358.
55. Earlier in *That Hideous Strength* (ch. 14, sec. 5, 317), Dr. Ransom had prophesied that once the N.I.C.E. people were no longer useful to the *eldila*, the *eldila* would destroy them.
56. *That Hideous Strength*, ch. 14, sec. 1, 296.

12

C. S. Lewis, Scientism, and the Battle of the Books

M. D. Aeschliman

Night Thoughts

Prussian idealism took the heart of flesh and blood from the German and in its place gave him one of iron and paper.

—Theodor Haecker (1940)[1]

For his open opposition to the German National-Socialist "New Order," the anti-Nazi humanist and writer Theodor Haecker (1889–1945) was prohibited from writing or speaking in public in his native Germany for the last ten years of his life. During these dark years (1935–1945), Haecker lived under house arrest and faced repeated raids and searches of his papers by the Gestapo.[2] He nevertheless secretly kept a journal of the years 1939–1945, during which his Munich home was destroyed by Allied bombing and he was gradually going blind. Haecker lived long enough to know that the "Thousand-Year Reich" would last only twelve years, but not long enough to see the surrender. His noble "night thoughts" were published in German only after his death.

During these same years, C. S. Lewis was living and teaching in Oxford and meditating on the epistemological, ethical, and educational issues that would lead to his Riddell Lectures of 1943, published as *The Abolition of Man*, which John West rightly calls "the best defense of natural law to be published in the twentieth century."[3]

Both Haecker and Lewis were philosophically literate, and both meditated on how it had become possible that Western civilization, for 150 years increasingly confident about its future in so many fields of activity, had collapsed into the nightmarish cauldron of history through which they were living. Both men came to believe that the roots of the satanic developments of their time were ultimately epistemological. Not found in the apparently primitive, superstitious, poor parts of the world that the Western nations had increasingly colonized, dominated, and exploited, these satanic modern developments were inventions of Western "civilization" itself, radiating out from its historic centers of high culture—France, Germany, Italy, England.

In his secretly written "journal in the night," Haecker observed with increasing gloom that the Germany whose education system, scientific research, industry, and technological accomplishments had become the envy and model of the "developed" world, had resurrected in modern garb the most immoral kinds of ethnocentrism and national egotism in the name of Darwinian "racial science."[4] The "race" that had produced Bach, Handel, Haydn, Mozart, Beethoven, Brahms, and Wagner, also produced an increasingly dehumanized modern intellectual tradition that had for 175 years grown into the most elaborate and prestigious forms of modern philosophy, from Kant, through Hegel and Marx and Nietzsche, to Freud and Heidegger. No more imposing systematic intellectual structure had—or has—ever been created in the history of the world.[5]

Having studied at the University of Berlin, Haecker was a learned beneficiary of this legacy, but he came to believe that its ultimate effect on Germany was catastrophic, and that the desertion of the classical epistemology and ethics first articulated in the works of Plato and Aristotle had played a large role in the tragedy of modern German history. "It seems as though the Germans had chosen madmen, men who quite simply went mad, and consecrated them, raised men like Nietzsche and Holderlin to the level of prophets, heroes, saints, and wise men, and made idols of them," he wrote in 1943.[6]

Like T. S. Eliot and C. S. Lewis across the English Channel, and Jacques Maritain, now living in America, Haecker hoped and prayed for German defeat during those terrible years 1939–1945. Also like Eliot, Lewis, and Maritain, Haecker believed the profound modern confusion—"the dismemberment and disintegration of the modern person" (1942 entry)—had epistemological, philosophical roots.[7] The supposedly enlightened and emancipated modern person "has a nihilistic, devastating philosophy once away from the privileged philosophy of being of Aristotle and Plato; he has a nihilistic politics, an apostate politics" (1943 entry).[8] Haecker lamented the "gnosticism and 'idealism' in German philosophy," with the latter "after all only a sort of watered-down gnosticism."[9]

The revenge against this German "idealism" was well under way in Haecker's and Lewis's life-time in equally barbaric, opposite, reductionistic extremes—Marxist "scientific socialism" in the Soviet Union; and the materialistic "logical positivism" of the "Vienna Circle" and A. J. Ayer: "the consequences deduced from the most threadbare 'scientific' hypothesis are looked upon as though they were eternal truths" (1940 entry).[10] Very few major nineteenth-century writers had foreseen the apocalyptic effects of these epistemological heresies. The far-seeing few included Kierkegaard, Newman, and Dostoevsky, to whom Haecker attributed "the spirit of prophecy."[11]

Although Haecker translated both Kierkegaard and Newman into German—the former so much influenced by Socrates, the latter by Aristotle—he also was in possession of "the privileged philosophy of being of Aristotle and Plato" and aware of its civilizing, deepening continuity—its momentum, trajectory, and effects across time—in the work of the "magnanimous" Virgil (who "would today be silenced in a concentration camp" (1940 entry))[12] and in the writings of St. Thomas Aquinas.

Sometimes in history, Haecker wrote in 1941, the difference and tension between a "Master" and a "schoolmaster" is overcome: "sometimes they are united in a single person, and that is the glory of the West. Sometimes the perfect 'Master' is the 'master of the Schools', the perfect schoolmaster. The greatest example is St. Thomas Aquinas" (1941 entry).[13]

It is no exaggeration to say that in our time that "perfect schoolmaster," an instance and revelation of the true "glory of the West," is C. S. Lewis.

In the perennial "battle of the books," the culture struggle or "kulturkampf" in educational policy, school curricula, the arts, the sciences, the universities, the popular culture, and "cultural politics," Lewis may be said to perform a role like that of Plato or Aristotle, Virgil, or St. Thomas—in part because their patrimony is vivid in his own writings.[14]

Surely over the last century there have been many other noble intellectual witnesses to the indispensable truths that humanize us: G. K. Chesterton, Haecker himself, T. S. Eliot (who drew on Haecker's work),[15] Jacques Maritain, Martin Buber, Reinhold Niebuhr, Richard Weaver, Leo Strauss, Michael Polanyi, Dorothy Sayers, Hans Jonas, Viktor Frankl, Philip Rieff, Stanley Jaki, Gertrude Himmelfarb, Jacques Barzun, Alasdair MacIntyre, and several recent Popes, to name a few of the most important. But Lewis's capacity to write about the most important issues of modern life and culture, in several genres and in a way that is not only intelligible but inviting and persuasive to many sorts of people, remains unequalled.

The Image of the Human Person

> [T]hat inextricable connection between the non-material self and freedom is the defining feature of man's exceptionality, for we, unlike our primate cousins, are free to forge our own destinies [so that each of us may] become that distinct, unique person responsible for his actions of which all human societies are composed, and from which virtually everything we value flows.
>
> —Dr. James LeFanu[16]

With the outcome of the great World War still in doubt in 1943, C. S. Lewis nevertheless chose not to attack directly either German Nazism or Russian "scientific-socialist" Communism in *The Abolition of Man*. Instead, he looked closely at influential epistemological, ethical, and educational ideas that had taken root throughout the "advanced" world in various but related, cognate forms.

Like Haecker, Lewis built on "the privileged philosophy of being of Aristotle and Plato," what Haecker also called, following his great countryman Leibniz, the "perennial philosophy" (*philosophia perennis*). Lewis recognized that only that philosophy and its subsequent trajectory did adequate justice to the nature of being itself.[17] By contrast, both Social-Darwinist Nazism/Fascism and "scientific-socialist" Communism offered grossly reductionist accounts of natural life, history, and the human person.

Reductionist scientific materialism, i.e., scientism, "entered the political field most significantly in the wake of the French Revolution of 1789," writes Tzvetan Todorov, Director of Research at the French Centre National de la Recherche Scientifique in Paris. "Such was the prestige of science throughout the 19th century that its cult flourished among all kinds of thinkers—both friends and foes of the Revolution— who all hoped to replace a collapsing religion with the rule of science." The chapter of Todorov's recent book in which he makes this analysis is entitled "What Went Wrong in the Twentieth Century."[18]

Lewis knew that the method of the natural sciences that had brought great technological triumphs (as well as disasters) to Western nations since the time of Francis Bacon in the seventeenth century was reductionist: It reduced the field of vision experimentally and theoretically to what could be measured and evaluated in terms of quantity, matter, space, time, mass, or energy. Thoughtful and well-informed observers of the growth of the natural sciences such as the American psychologist William James (d. 1910) had hoped and worked for a "radical empiricism" that would include and not eliminate human phenomena such as purposive behavior, free will, initiative, truth, morality, reason, and the laws of logic (the principle of non-contradiction, the syllogism, rational inference), which were themselves the indispensable presuppositions of the scientific method itself.[19] But the "radical empiricism" that developed throughout the nineteenth century increasingly eviscerated its own human and rational basis, fulfilling the Renaissance French humanist Rabelais's fear that "science without conscience is the death of the soul," the elimination of the awareness and the reality of the human subject who has undertaken the scientific enterprise in the first place.[20]

The great contemporary historian of modern Germany Fritz Stern—as a boy himself a German-Jewish refugee from the Nazis and subsequently Provost of Columbia University—wrote of the intellectual milieu of "dogmatic scientism" in late nineteenth- and early twentieth-century Germany in which Theodor Haecker grew up:

> the enormously successful [scientific] popularizers Buchner, Moleschott, and [Ernst] Haeckel... threw all scientific caution to the winds, and, generalizing from [experimental] beginnings, preached a scientific materialism that was more comprehensive but no less crude than that of some of the late 18th-century [mainly French] philosophers. In this view of the universe, only matter—that is, the atom in its infinite combinations—had reality; the human mind and will as independent forces were condemned as figments of the "idealistic" imagination.[21]

Lewis's case against the scientific materialists was an instance of the perennial rational case against all forms of rational self-contradiction. Logical consistency has been the main warrant for rationality and elementary self-knowledge ever since Socrates had modeled it in Plato's dialogues and Aristotle had discussed it in terms of the principle of non-contradiction in the *Metaphysics* (1005b19–20; 1006a1). And Lewis knew that there were contemporary scientists and natural philosophers who recognized the thoroughly discrediting character of this scientistic self-contradiction, such as his contemporary A. N. Whitehead (1861–1947). One of Whitehead's lapidary insights was that scientists "animated by the purpose of proving themselves purposeless constitute an interesting subject for study."[22]

Whitehead's point is a classic instance of indicating a thoroughly discrediting thematic-performative self-contradiction: No assertion can be true if it exempts its own spokesman from the generalization that he is himself making. Insisting that language and action cannot really be rational, purposive, or effectual—a standard, recurrent scientific-materialist, reductionist rhetorical-logical move—is itself a self-contradiction because the speaker has himself just made such a purposeful, allegedly rational speech act, which also assumes that it can have an effect on the hearer or reader, and the hearer or reader may choose to believe it or not believe it on the basis of its rationality or "warranted assertability." It was with such a rational tactic that Socrates critiqued and defeated the immoralist Thrasymachus in *The Republic* of Plato, and ever since the Platonic dialogues it has been part of the common currency of rationality in the West, along with the syllogism or rational inference.[23]

Lewis believed that the history of modern culture and education had brought both gain and loss. How could any sensitive observer doubt the latter, given the indubitably Luciferian character of much world history during Lewis's lifetime (1898–1963)? Modern natural science had brought many great goods—most obviously in the fields of medicine and agriculture—but it had purchased them at very high cost, not only to the physical environment but to the understanding of the human person as a rational and moral agent. "Modern [scientific] theory is about objects lower than man," Hans Jonas has written.[24] "For a scientific theory of him to be possible," the human person, "including his habits of valuation, has to be taken to be determined by causal laws, as an instance and part of nature." And "the scientist does take" the human person to be so determined by causal laws, "but not himself while he assumes and exercises his freedom of inquiry and his openness to reason, evidence, and truth."[25] Thus we have a classic thematic-performative self-contradiction of the kind Michael Polanyi, a distinguished physical chemist, also never ceased to point out and protest.[26]

"Man-the-knower," Jonas continues, "apprehends" his human object as something lower than himself, as a sub-human or merely natural object, "since all scientific theory is of things lower than man-the-knower." But it is only "on that condition" that human objects "can be subject to 'theory,' hence to control, hence to use. Then man-lower-than-man," the human object stripped of subjectivity and human status, as "explained by the human sciences—man reified—can by the instructions of these sciences be controlled (even 'engineered') and thus used."[27]

Jonas and Polanyi were both Jewish refugees from Nazi Social Darwinism. Theodor Haecker had noted in his 1941 night journal that Jews in Germany had begun to be required to wear the star of David on their coats.[28] Even by 1943 it is unlikely that he or Lewis, delivering the Riddell Lectures that became *The Abolition of Man*, can have imagined the extent to which scientistic, eugenic, dehumanizing Nazi beliefs had been institutionally, industrially arranged and "engineered" so as to liquidate large numbers of people—Jews, Slavs, and the "unfit"—whose human status had first been mentally and theoretically liquidated by scientistic ideology in the way Jonas describes. Similar thinking about "class enemies" among the Communists to the East had already led to the "purging" or elimination of whole categories of people, deprived by Marxist "scientific socialism" of human status and worth.[29] Ideas have consequences: Here scientistic ideas had lethal ones.

Lewis was probably unaware of the Nazi Holocaust already underway in 1943, but nevertheless, in the final chapter of *The Abolition of Man*, he considered that the apparently wonder-working development of modern science and technology—nowhere more apparent than in Germany and Britain—may have been bought at too high a price. "Little did he know," we may say.

Writing much more recently, the distinguished Jewish-American geneticist Robert Pollack has commented that the "vision of the natural world" that modern, exclusively scientific "explanations produce is simply too terrifying and depressing to me to be borne without the emotional buffer of my own religion."[30] Rabelais had prophetically warned nearly 500 years before that "science without conscience is the death of the soul," but it took the twentieth century to document these effects in gruesome detail.

Language and Humanity

> The only way to avoid metaphysics is to say nothing.
> —E. A. Burtt[31]

According to Ian Tattersall, Curator of the American Museum of Natural History, "All the major intellectual attributes of modern man are tied up in some way with language."[32] Yet ever since the seventeenth century, a powerful scientistic current has increasingly attempted to "demystify" language and assimilate it without remainder to the physical, material, or "natural" realm. The British Royal Society, the first great modern scientific affiliation, was founded in 1660 with a Latin motto from Horace, *nullius in verba*—"nothing in words," which of course is a thematic-performative self-contradiction, since this assertion of the uselessness and ineffectuality of words is here put in *words*. The goal would be a fully mathematical language of numerals and quantities that describe objects and material processes. Jonathan Swift had satirical fun depicting this project and its self-contradictions in Book Three of *Gulliver's Travels* (1726).[33]

For Lewis, the high cost of "methodological reductionism" or "methodological naturalism" was not only what it does to nature, though as he said in *The Abolition of Man*, a "regenerate science" would "not do even to minerals and vegetables what modern science threatens to do to man himself." Nor was Lewis only concerned with the spectacular iniquity that after 1945 in Germany and after 1976 in Russia the world came to know had been done in those nations and for which the names "Auschwitz" and the "Gulag Archipelago" are now permanent markers.

Instead, Lewis was primarily concerned with the elimination of ethics as an entire *category* of thought by thoroughgoing materialistic "logical positivists" such as Rudolf Carnap and Lewis's own Oxford colleague A. J. Ayer.[34] In America, the career of B. F. Skinner, the "behavioral" psychologist at Harvard, was another illustration of how the program of "methodological reductionism" or "methodological naturalism" or "methodological materialism"—supposedly only a heuristic program—almost inevitably slides, by a kind of nihilistic gravity, into the ultimate stance of "ontological materialism," an ultimate denial of truth, free will, ethics, purpose, the laws of logic, the integrity of language, and the identity and dignity of the human person.

Skinner wrote a book called *Beyond Freedom and Dignity*, though one assumes that he thought that he had the freedom to write it and others the freedom to read it (or not). In a mordant comment, the philosopher Sidney Morgenbesser reacted to Skinner's dehumanizing project by wittily remarking that Skinner thinks "[w]e shouldn't anthropomorphize people," which is to say, conceive and treat them like human beings, with separate autonomous personalities, mutual rational interests, and reciprocal moral obligations.[35]

The American philosopher of science Nancey Murphy has recently said that a religion or philosophy of scientific naturalism "ought not to be allowed *establishment* in the curriculum of the [American] public schools,"[36] but this was one emergent development that Lewis noticed in Britain by 1943, which is why his subtitle to *The Abolition of Man* explicitly concerns educational policy: "reflections on education with special reference to the teaching of English in the upper forms of schools."

Twelve years later, in 1955, the distinguished American philosopher of education Philip H. Phenix noted that the "secularization" of the American public schools had "to some extent become a method of propagating a particular dogmatic faith, namely, scientific naturalism."[37] And it was largely scientific naturalism, with its incapacity to recognize and legitimate the autonomy of language, free will, rationality, or ethics, that has led to the varieties of decadence and barbarism that have characterized the West in the twentieth century. As the American political philosopher Harry Jaffa has written, "[t]he doctrine that man, the intelligent being, is 'caused' by an unintelligent first principle cannot... escape self-contradiction. Intelligence is an irreducible reality."[38] And its vehicle is language.

The Battle of the Books

> When all that says 'it is good' has been debunked, what says 'I want' remains. It cannot be exploded or 'seen through' because it never had any pretensions [to virtue, justice, or ethics].
> —C. S. Lewis, *The Abolition of Man*, chapter 3

Lewis was not only a trained philosopher, literary historian, and literary critic; he was a versatile imaginative and expository writer of great power. As a philosopher and literary historian he knew that wherever and whenever in the history of the world there had been anything like what we would call a "high" civilization—with all its ambiguities—there had been a "battle of the books." He knew that his argument, above quoted, about the debunking of all moral assertions was the same argument made by Socrates against Thrasymachus, and implicitly against the dominant values of Athens in his time, in Plato's *Republic*. Rebelling against the "perennial philosophy," and apparently triumphing over it with spectacular signs and wonders, modern scientific naturalism had reached an impasse, a vertiginous free fall into rational and moral nihilism. "Whenever anyone says anything is good, all he is saying is that he likes it" was the upshot of A. J. Ayer's logical-positivist "emotivism," sometimes called the "booh!-hurrah!" theory of ethics; the apocalyptic 1930s seems a particularly unpropitious time for a philosopher to toy with such immoralism.

Bertrand Russell also put it quite shamelessly: "The theory which I [advocate] is a form of the doctrine which is called the 'subjectivity' of values, [which] consists in maintaining that, if two men differ about values, there is not a disagreement as to any kind of truth, but a difference of taste." Russell hardly shied away from the aestheticist, hedonist, immoralist implications of this view, despite the fact that he was a fiercely verbal public moralist all of his life. "If one man says 'oysters are good' and another says 'I think they are bad,' we recognize that there is nothing to argue about. [My subjectivist] theory," Russell concluded, "holds that all differences as to values are of this sort, although we do not naturally think them so when we are dealing with matters that seem to us more exalted than oysters."[39] Such ironically debunked "exalted matters" include, of course, ethics, law, and politics. The volubly self-righteous Russell provides a classic illustration of what Michael Polanyi called "moral inversion"—a fierce, Pharisaical moralism whose own naturalistic philosophy has actually destroyed the rational grounds for any and all ethical obligation per se.[40]

Russell and the other Bloomsbury-Cambridge aesthetes and Leftists helped create the "climate of opinion" that Lewis critiques in *The Abolition of Man*, "the whole system of values which happened to be in vogue among moderately educated young men of the professional classes during the period between the two [world] wars."[41] Writing retrospectively in 1938, Russell's Bloomsbury-Cambridge friend John Maynard Keynes described their own clever "immoralism": "Our comments on life and affairs were bright and amusing, but brittle… because there was no solid diagnosis of human nature underlying them. Bertie [Russell] in particular sustained simultaneously… opinions ludicrously incompatible." Keynes goes on in his remarkably revealing memoir to say of his "set": "[We expanded] illegitimately the field of aesthetic appreciation [by] classifying as aesthetic experience what is really human experience and somehow [sterilized] it by this mis-classification." Any imaginable good (e.g., justice) thus becomes a merely personal preference or taste, producing, in Keynes's own words, "libertinism and comprehensive irreverence." Furthermore: "We repudiated entirely customary morals, conventions and traditional wisdom. We were, that is to say, in the strict sense of the term, immoralists… we recognized no moral obligation on us, no inner sanction, to conform or to obey."[42] It was from this "climate of opinion" that the highly privileged Cambridge Communist spies Burgess, Maclean, Philby, and Blunt came, what Andrew Boyle called *The Climate of Treason*.[43] We "laugh at honour," Lewis wrote at the end of chapter 1 of *Abolition*, "and are shocked to find traitors in our midst."[44]

Writing about Russell in his diary for 1950, Malcolm Muggeridge characterized "his kind of mind and point of view, though [initially] attractive, [as] the most ruinous of all, the most destructive, the most cowardly, the most lamentable. His reason tells him to seek what pleases and to shun the disagreeable, and, obeying it, he becomes utterly selfish, indifferent to all forms of responsibility, personal and social."[45]

As we can see, the era since August 1914 has been disfigured by several forms of human vice, folly, and fecklessness, many of them conveyed by "scientistic" or aesthete intellectuals, with numerous examples of "learned foolishness," and some of intellectual iniquity, conducing to self-contradiction, demoralization, dehumanization, destruction, and tragedy.

By contrast, C. S. Lewis was a great, luminous, orthodox scribe, and an antiseptic "guide for the perplexed," one who bowed to the constraint of truth, and thus climbed its ladder.

Endnotes

1. Theodor Haecker, *Journal in the Night*, translated from German by Alexander Dru (New York: Pantheon, 1950), 72.
2. See Alexander Dru, introduction to Haecker, *Journal*.
3. See West's entry on *The Abolition of Man* in *The C. S. Lewis Reader's Encyclopedia*, ed. J. D. Schultz and J. G. West (Grand Rapids, MI: Zondervan, 1998), 67–69, and my own *The Restitution of Man: C. S. Lewis and the Case Against Scientism*, second edition (Grand Rapids, MI: Eerdmans, 1998). Also see C. S. Lewis, *The Abolition of Man* (New York: Touchstone, 1996).
4. Haecker: "Inequality in the place of equality, for the whole movement goes back to an essay by Gobineau on the Inequality of Races." 1940 entry, Haecker, *Journal*, 58. See Richard Weikart, *From Darwin to Hitler: Evolutionary Ethics, Eugenics, and Racism in Germany* (New York: Palgrave Macmillan, 2004). For the widespread admiration of Prussian and then German Imperial public schooling in the 19th century, see Charles L. Glenn, Jr., *Contrasting Models of State and School: A Comparative Study of Parental Choice and State Control* (New York: Continuum, 2011).
5. See, e.g., Julian Roberts, *German Philosophy: An Introduction* (Cambridge, UK: Polity Press, 1988); Jurgen Habermas, *Philosophical-Political Profiles*, trans. F. G. Lawrence (Cambridge, MA: MIT Press, 1993).
6. Haecker, *Journal*, 202.
7. Ibid., 178. After listening to Goebbels on the radio in 1940, Haecker wrote: "The religion of the German *Herrgott* is the religion of the 'heart of stone'. They will be beaten, they will be ground to powder, and then they will once again desire a heart of flesh and blood." Ibid., 32.
8. Ibid., 203.
9. Ibid., 202–203.

10. Ibid., 43.
11. Ibid., 171, 169–170. Haecker's translator Dru rightly adds the historian Burckhardt to the list of prophets. Ibid., xxxvii ff.
12. Ibid., 59.
13. Ibid., 152.
14. Lewis translated much of Virgil's *Aeneid* and his translation has been recently published. See A. T. Reyes, ed., *C. S. Lewis's Lost "Aeneid"* (New Haven, CT: Yale University Press, 2011).
15. See Eliot, "Virgil and the Christian World" (1951), in *On Poetry and Poets* (London: Faber and Faber, 1957), especially 123 and following pages.
16. James LeFanu, *Why Us? How Science Rediscovered the Mystery of Ourselves* (New York: Vintage/Random House, 2011), 254–255.
17. For Haecker's self-identification as a "disciple of the *philosophia perennis*," see the 1943 entry, #639, Haecker, *Journal*, 197. On Leibniz's use of the term "perennial philosophy," see Aldous Huxley, *The Perennial Philosophy* (London: Fontana, 1958), 172.
18. Todorov, *Hope and Memory: Reflections on the Twentieth Century*, English translation (Princeton: Princeton University Press, 2003), 23. Cf. Aeschliman on the Frenchman Ernst Renan's "religion of science," Aeschliman, *The Restitution of Man*, 36.
19. See William James, *Essays in Radical Empiricism* (New York: Longmans, Green, 1912). For an excellent, brief, recent antidote to reductive ideas of nature, see Christopher O. Blum, "Mirrors Held Up to Nature," *Modern Age*, 52, no. 3 (Summer 2010): 230–233.
20. For further discussion and documentation, see Aeschliman, *The Restitution of Man*, chapter 2 and passim. See also Michael Polanyi on the effects of "the programme of comprehensive doubt" and its denial of "the fiduciary rootedness of all rationality," eventuating in "modern fanaticism," from whose Nazi form Polanyi himself was a refugee. Polanyi, *Personal Knowledge: Towards a Post-Critical Philosophy* (New York: Harper Torchbook, 1964), 297–298, 308–309.
21. Fritz Stern, *The Politics of Cultural Despair* (Garden City, NY: Doubleday Anchor, 1965), 161–162.
22. Whitehead, *The Function of Reason* (Princeton: Princeton University Press, 1929), 12. A careful defense and extension of Lewis's "foundationalist" or "logocentric" philosophical position is Victor Reppert's *C. S. Lewis's Dangerous Idea: In Defense of the Argument from Reason* (Downers Grove, IL: InterVarsity Press, 2003).
23. See Ben F. Meyer, "The Philosophical Crusher," *First Things* (April 1991): 9–11. Victor Reppert's book (previous note) is a good introduction to this set of issues, as is the work of Michael Polanyi (note 20). On the evacuation, evisceration, occlusion, or reification of the human subject in much modern thought, see also Hans Jonas, *The Phenomenon of Man: Toward a Philosophical Biology* (New York: Harper and Row, 1966), 196; and Aeschliman, *The Restitution of Man*, 55.
24. Jonas, *The Phenomenon of Man*, 195.
25. Ibid., 196.
26. On Polanyi, see note 20 above and M. D. Aeschliman, "Saving Remnant," *National Review* (February 12, 2007): 45–46.
27. Jonas, *The Phenomenon of Man*, 196.

28. Entry # 583, Haecker, *Night Journal*, 173. Haecker imagined that in a future time Germans abroad should be obliged to wear the swastika, the "hook-cross" (*Haken-kreuz*), on their coats, as "the sign of the anti-christ."
29. See, e.g., Richard Pipes, *Communism: A History* (New York: Modern Library, 2001); Stephane Courtois et al., eds., *The Black Book of Communism*, English translation (Cambridge, MA: Harvard University Press, 1999); M. D. Aeschliman, Introduction to Malcolm Muggeridge, *Winter in Moscow* (Grand Rapids, MI: Eerdmans, 1987). Established Western intellectuals still seem far from eager to concentrate on the degree to which "progressive" intellectual sympathy and collaboration with Communism encouraged its growth and consolidation in Russia, eastern Europe, and the Far East.
30. Robert Pollack, *The Faith of Biology and the Biology of Faith* (2000), quoted in *Research News and Opportunities in Science and Theology*, 3, no. 4 (December 2002): 16.
31. *The Metaphysical Foundations of Modern Physical Science* (1932 ed.), quoted by Stanley L. Jaki, "The Demythologization of Science," in *"The Absolute beneath the Relative" and Other Essays* (Lanham, MD: University Press of America, 1988), 201. See also my article "Science and Scientism" in *American Conservatism: An Encyclopedia*, ed. Bruce Frohnen et al. (Wilmington, DE: ISI Books, 2006), 771–775.
32. Tattersall, quoted by James LeFanu, *Why Us?*, 50. See LeFanu's subsequent discussion of language, and the importance of Noam Chomsky's work on it, 50–56. And cf. Chomsky's epistemological chastity: "As soon as questions of will or decision or reason or choice of action arise, human science is at a loss." *The Listener* (London), April 6, 1978.
33. On such issues, Allan Bloom has written, "Swift's perspicacity is astonishing." "Giants and Dwarfs: An Outline of *Gulliver's Travels*," in *Giants and Dwarfs: Essays 1960–1990* (New York: Touchstone, 1990), M. Horkheimer and T. Adorno argue that "number became the canon of the Enlightenment." *Dialectic of Enlightenment* (New York: 1994), 7. An invaluable analysis.
34. For detail and discussion, see Aeschliman, *The Restitution of Man*, chapter 4.
35. Quoted by John Silber, "Science versus scientism," *The New Criterion* (November 2005): 12.
36. Murphey's comment can be found in her 1993 essay "Phillip Johnson on Trial," in *Intelligent Design Creationism and Its Critics: Philosophical, Theological, and Scientific Perspectives*, ed. Robert T. Pennock (Cambridge, MA: MIT Press, 2001), 464.
37. *Teachers College Record*, 57 (October 1955): 30.
38. Harry V. Jaffa, "In Defense of the 'Natural Law Thesis,'" in *Equality and Liberty: Theory and Practice in American Politics* (New York: Oxford University Press, 1965), 200.
39. Russell, "Science and Ethics" (1961), quoted in Reppert, *C. S. Lewis's Dangerous Idea*, 41–42.
40. On the contradictions and derelictions of Russell's thought, behavior, and character, see Malcolm Muggeridge's brilliant satirical review of Ronald Clark's biography of Russell, *The New Republic*, 174, no. 14 (April 3, 1976), which I think is one of the outstanding essays of the twentieth century in English, but not as yet re-printed. On "moral inversion," see Michael Polanyi, *Personal Knowledge*, 231–35; Mark T. Mitchell, *Michael Polanyi: The Art of Knowing* (Wilmington, DE: ISI Books, 2006), 52–58; and notes 20 and 25 above.

41. Lewis, *Abolition of Man*, chapter 2.
42. J. M. Keynes, "My Early Beliefs," in *Two Memoirs* (London: Rupert Hart-Davis, 1949), 102–103, 98.
43. Andrew Boyle, *Climate of Treason* (London: Hodder and Stoughton Coronet, 1980). See also Richard Deacon, *The Cambridge Apostles* (New York: Farrar, Straus, and Giroux, 1985); Goronwy Rees, *A Chapter of Accidents* (New York: Library Press, 1972); Malcolm Muggeridge, *Chronicles of Wasted Time, Vol. II: The Infernal Grove* (London: Collins, 1973), especially chapter 2, "Grinning Honour," and his essay "The Eclipse of the Gentleman," *Time* (December 3, 1979), 33–4, another great piece of twentieth-century writing.
44. Lewis, *The Abolition of Man*, 37.
45. Entry for October 15, 1950, in J. Bright-Holmes, ed., *Like It Was: A Selection from the Diaries of Malcolm Muggeridge* (London: Collins, 1981), 415.

13

C. S. Lewis, Scientism, and the Moral Imagination

Michael Matheson Miller

At the beginning of *The Abolition of Man*, C. S. Lewis comments, "I doubt whether we are sufficiently attentive to the importance of elementary text-books."[1] As he goes on to explain, by the time a student arrives at university or reads the editorial page much of the battle for his mind has already taken place. An orientation has been set. The intellectual and imaginative scaffolding of the way in which he will view the world has already been built. "It is not a theory they put in his mind, but an assumption, which ten years hence, its origin forgotten and its presence unconscious, will condition him to take one side in a controversy which he has never recognized as a controversy at all."[2]

One service Lewis offers his readers is to root up some of the more pernicious of these unexamined assumptions and hold them up to the light, beginning with scientism. Scientism, not to be confused with science or scientific investigation, is the view that empirical reality is the only reality and anything not empirical, i.e., not subject to measurement and analysis by the scientific method, is merely subjective or unimportant. Scientism is closely related to rationalism, which posits that only those things that are empirically demonstrable fall within the purview of reason. This means that all non-empirical reality is relegated to the non-rational. Thus all questions of morality, goodness, beauty, kindness, compassion, mercy, are no longer seen as being rational and become reduced to subjective opinion or feeling.

One of the key intellectual labors of Lewis's life, running through all of his work from his scholarly essays and Christian apologetics to his children's stories and science fiction, was a critique of scientism, which he believed led to relativism and, in the process, undermined the foundations of true science. More than this, Lewis was convinced that scientism and its poisonous fruit dulled the intellect and stole from people, often without their knowing, the most important experiences of their lives, including those things which make us fully human.

Scientism, Lewis argued, is incoherent at its root, and if followed to its illogical conclusion would lead to the end of Western civilization and the "abolition of man."

He was especially worried about how modern education based on debased reason and devoid of a pursuit of truth would strip away the most essentially humanizing elements and produce "Men without Chests,"[3] that is, an impoverished race of people divorced from their heritage. Lewis argued that scientism and rationalism make things like justice, nobility, honesty, valor, courage, and compassion non-rational and undermine the human capacity for greatness. Such thinking takes away the heart or what Plato called the "spirited element" of the person. It removes the "chest" and leaves man with only the passions and a corrupted intellect. This may not be the intent of scientism, but it is the result:

> [S]uch is the tragi-comedy of our situation—we continue to clamor for those very qualities we are rendering impossible. You can hardly open a periodical without coming across the statement that what our civilization needs is more 'drive,' more dynamism, or self-sacrifice, or 'creativity.' In a sort of ghastly simplicity we remove the organ and demand the function. We make men without chests and expect of them virtue and enterprise. We laugh at honor and are shocked to find traitors in our midst. We castrate and bid the geldings be fruitful.[4]

Lewis's image of "Men without Chests" evokes the warning of fellow Irishman Edmund Burke from two centuries prior about the intellectually denuding effects of the Enlightenment rationalism of the *philosophes*: "All the decent drapery of life is to be rudely torn off. All the superadded ideas, furnished from the wardrobe of a moral imagination, which the heart owns and the understanding ratifies as necessary to cover the defects of our naked, shivering nature, and to raise it to dignity in our own estimation, are to be exploded as a ridiculous, absurd, and antiquated fashion."[5]

Both men, in their time, sensed the creeping effects of a view that at its heart is irrational, or to come at in Chestertonian terms, is mad for having taken a partial truth about the human person and mistaking it for the whole truth.[6]

As a medicine for the madness, we could do worse than to use *The Abolition of Man* as a springboard into an exploration of how scientism affects the human spirit, how its truncated vision of reality cuts us off from the traditional ends of human seeking—the good, the true, and the beautiful. That is the aim of this chapter.

Just as the process of absorbing the assumptions of scientism takes place long before a young man or woman ever enters a university classroom, the effects of scientism go far beyond science into politics, the arts, and human relationships. Scientism is a form of reductionism, and Lewis, like Burke, recognized that what is needed to combat reductionism is clear, reasoned argument, a recovery of what Lewis called "reasonable" emotions[7] and a renewal of what Burke called the "moral imagination." The English historian and sociologist Christopher Dawson made a related point when he called for the recovery of the "lost channels of communication"[8] which train passions, sentiments and imagination as well as the intellect. An exploration of these ideas will, at the end of the essay, lead in to some reflections on how we might rebuild the moral imagination to help form men and women of clear reasoning and noble sentiments—"men *with* chests."

Elementary Education:
Building the Intellectual Scaffold of Scientism

As an intellectual position, scientism isn't what you might call robust. Even a cursory glance at it reveals the self-refuting nature of its relativism and the shaky foundations of its reductionism. Yet scientism is still attractive and widely held. Why? The answer is at least two-fold.

First, and as noted at the beginning of the essay, actual arguments for scientism are rarely given. Instead "assumptions are put in students' heads" that prepare the mind to accept scientism later. These assumptions dull the intellect and make it lazy. Second, and equally important, these assumptions are couched in an educational setting and a culture that increasingly stunts the development of the moral imagination.

In *The Abolition of Man* Lewis showed how a textbook he called *The Green Book* that was supposed to be teaching English instead indoctrinated the students into rationalism without the students' knowing what was happening. In a discussion on the thought and expression of Coleridge, controversial philosophical positions of moral and aesthetic relativism were put forth without argument. Yet it was precisely because there was no argument that it had such a more powerful effect on the student. The student would tend to regard the discussion as an objective lesson about language and meaning. As Lewis writes, the student was focusing on his grammar and not realizing what was at stake philosophically. The power of the argument rests in the fact that they are "dealing with a boy: a boy who thinks he is 'doing' his 'English prep' and has no notion that ethics, theology and politics are all at stake. It is not a theory they put into his mind but an assumption… which will condition him to take one side in a controversy which he has never recognized as a controversy at all."[9]

Today there is a highly charged battle about the culture of schools. Violence, poor performance, a morally relativized form of sex-education, advocacy for homosexual marriage, and a moral relativism masquerading as tolerance and diversity dominate the school systems. Many parents, in order to spare their children from this "brave new world," spend thousands and tens of thousands of dollars to send their children to private schools or, increasingly, educate their children at home. While this can protect children from a good deal of cultural indoctrination, many of the intellectual influences of scientism and its relativism remain. As Mark Roberts, has argued,[10] even in these private, Christian schools and homeschools, we often are not "sufficiently attentive to the importance of elementary text-books."[11] Too often we reduce Christian education to modern secular education with Jesus sprinkled on top, or classical education to modern education with Latin and Greek. Roberts analyzed homeschool textbooks and shows how the intellectual error of rationalism was subtly introduced, not by argument but through a seemingly harmless, critical thinking exercise.[12]

The child is given a list of several statements and has to identify which statements are "facts" and which are "opinions." You may remember it from your own schooldays.

The exercise goes something like this:

- Mozart was born in Salzburg.
- Mozart wrote beautiful music.
- John Paul II was the Pope for over 20 years.
- John Paul II was a good Pope.
- Bell-bottoms were popular in the 1970s.
- Bell-bottoms are cool.

Before getting to the main issue, it is worth noting that the exercise does not even make sense. The difference between fact and opinion is much more complex than simple. We should think more critically about the exercise itself.

The real problem is "the assumptions that are put into the child's head." The child is taught exactly what Lewis discussed in "Men without Chests"—that all value judgments are "*only...* something about our own feelings"[13]—and that only empirically verifiable statements are important. In other words, whatever is true, beautiful, just, noble, kind compassionate, or merciful is unconsciously determined to be "subjective and trivial." The entire Western tradition, in fact human tradition, is turned on its head in a "critical thinking" exercise. In one fell swoop the child is cut off from his patrimony without even knowing it, and a certain intellectual smugness is introduced. "[S]ome pleasure in their own knowingness will have entered their minds," Lewis writes, and one more "little portion of the human heritage has been quietly taken from them before they were old enough to understand."[14]

The exercise introduces aesthetic and moral relativism so innocently. After all, who would disagree that musical taste is a matter of opinion? Yes, musical taste may be a matter of opinion, but Mozart's music is beautiful whether we happen to like it or not. John Paul II was indeed a good pope, and notwithstanding the freedom to have bad taste or nostalgia for disco music, bell-bottoms are not cool. Too often we conflate our preferences with objective reality. But sometimes those preferences may in fact reflect a deeper reality.

In the critical thinking exercise, scientism and its reductionist rationalist worldview are presented as normal and non-controversial. The child can only conclude that all values are just subjective opinions. And from this, of course, moral and value relativism logically follow. The child is given the orientation toward relativism and is unaware of what is taking place.

The most important part of the exercise is actually what is *not* included. The example is usually related to art, history, or fashion. What if the following statements had been included?

- Murder is the intentional killing of an innocent person.
- Murder is bad.

I have done this exercise numerous times and have seen faces drop when I add the last one because all of a sudden the person realizes the implications of consigning all matters of value to the realm of opinion.

In that moment, my hope is that students will recognize that what is being fobbed off on them as rigorous thinking is actually sloppy and lazy thinking. First, it implies that to think critically one must be an empiricist, which is anything but critical, and second—and equally bad, it produces "Men without Chests," men unable to clearly and simply denounce murder as bad. This is not critical thinking or serious intellectual analysis. Lewis writes that "it is an outrage that they should be commonly spoken of as Intellectuals... It is not excess of thought but defect of fertile and generous emotion that marks them out. Their heads are no bigger than ordinary: it is the atrophy of the chest beneath that makes them seem so."[15] Emotional life and the life of the mind are severed completely. Any question about whether something is good, just, true, or beautiful no longer has to be wrestled with. It is merely an opinion—a feeling.

Before the young student even reaches high school, all the most important human and social questions that he will face throughout his life are relegated to the realm of subjective preference. Justice, truth, beauty, goodness, right, wrong, kindness, compassion, hope, fear, sadness, joy, love, friendship, wonder and so on. All of them are just matters of opinion—none of them is rational. All of this is done without any real argument. Entire traditions of thinking that shaped classical and Christian learning for 2000 years are rejected without any comment. They are not refuted; they are simply ignored. Later in life when he is offered chemical, genetic, or biological explanations for his emotions he will be intellectually prepared to accept them, never realizing why such explanations, so contrary to the way he actually experiences life, sound plausible. The intellectual scaffolding of scientism was set up somewhere around fourth grade. His intellect has been dulled and his humanity impoverished.

A so-called critical thinking exercise has removed, "long before he is old enough to choose, the possibility of having certain experiences which thinkers of more authority… have held to be generous, fruitful, and humane."[16] The young student's heritage is taken from him, and he didn't even realize it.

The Effects of Scientism on Human Experience

ALL OF this has profound consequences for political and social life. Lewis was clear that the problem of scientism was not merely limited to science or to those with academic interests. It fundamentally alters the way that people deal with the aspects of life that make us most human—truth, beauty, and goodness. The hobbled rationalism of scientism vitiates human experience, narrows our horizons, and makes everything from authentic politics to education, love, and friendship impossible since all of these things are focused on non-empirical goods. Lewis recognized the need to expand reason and, with this, recapture the idea of reasonable emotions without which we cannot live an integrated life in politics, education, the arts, or even our personal experiences.

Politics: Reason or Thrasymachus?

Plato believed the polis is "the soul writ large," and one of the most evident ways to see the effects of scientism on the human soul is to first see it from the broad perspective of politics.

When put into practice, scientism makes politics impossible because politics is not primarily about social organization, but about justice. As Pope Benedict argued before becoming Pope: "Politics is the realm of Reason, not of a merely technological, calculating reason, but of moral reason, since the goal of the state, and hence, the ultimate goal of politics, has a moral nature, namely peace and justice."[17]

In a framework of scientism where reason is reduced to the empirical, reasoned discourse about justice is no longer possible. All questions of justice and morality are relegated to the realm of subjective opinion and emotion or utility and a cost/benefit analysis. Either way, we are returned to Plato's bad guy in the *Republic*, Thrasymachus who argues that justice is merely the right of the stronger: Power equals truth, might makes right—or in our situation consensus, and increasingly, technology and bureaucracy equal truth.

Yet technology cannot tell us what we ought to do, and consensus is not always rational or just. In fact, it is often a crude combination of pragmatism and emotivism. In the words of Pope Benedict: "[T]he majority cannot be an ultimate principle since there are values that no majority is entitled to annul. It can never be right to kill innocent persons, and no power can make this legitimate. Here too, what is ultimately at stake is the defense of reason. Reason—that is moral reason—is above the majority."[18]

Notice Benedict's point. What it at stake is "the defense of reason." Unless we rehabilitate reason, we cannot have genuine politics, and this makes it very dangerous for minorities and the weak and defenseless. A utilitarian calculus does not leave much room for them, and their lives become dependent on the will and whims of the majority. It also tears asunder the foundations of democratic life and a free society.

Democracy, or more properly, representative government, requires rational discourse where all of the participants, even in the midst of disagreements over policy, at least agree that truth exists, that justice is better than injustice, and freedom is better than non-freedom, and that we can have a rational discussion about the best way to achieve these social, political, and human goods. Once scientism-fed rationalism becomes the norm, representative government is reduced to populism, or in many countries truth becomes determined not by a dictator, but by an amoral technocratic elite, similar to what the late Samuel Huntington called "Davos Man." Davos Man is not so far off the characters of Lord Feverstone and John Wither in *That Hideous Strength*—Lewis's fantasy dramatization of *The Abolition of Man*.[19] In Lewis's novel, these men were acting as the "conditioners" whose minds were freed from convention and the limiting "doctrine of objective values" to create a new world.

Scientism cannot be a foundation for a free, moral, and just society. It may sound alarmist but it is only stating a fact based on the claims of scientism itself. By rendering reasoned discourse about justice and right and wrong impossible, scientism leaves only two choices—raw power, or some pseudo-scientific decision-making tool, which Lewis notes will be held by those in power. One need only look at the rationalist political experiments of the French Revolution, the Soviet Union, German National Socialism, or Mao's cultural revolution to see the results of such an outlook when taken to its logical conclusion. All too often these regimes are portrayed as historical aberrations of madmen or ideologues. True, their leaders were madmen and ideologues, but what official retellings of these tragic historical episodes often fail to convey is that they were madmen and ideologues in no small measure *because* they followed scientism to its logical conclusion.

Recognizing this helps us understand that the threat of such madness remains. Under a rubric of scientism, our level of technology only determines what we *can* do; it does not give us direction for what we *ought* to do. Because morality is non-empirical, there is no reasonable means for us determine what is a moral or immoral use of technology. Instead, this must be determined by utility and pleasure, and therefore by whoever is in power.

This makes life difficult for minorities, the sick, infirm, and those who lack a voice. It is increasingly evident in Western societies that technology without morality is not always beneficial for every group—this is clearly the case for the unborn and increasingly, the elderly. Politics is "in the realm of reason," and when reason is corrupted politics is reduced to power. As Lewis writes in *The Abolition of Man*: "What we call Man's power is, in reality, a power possessed by some men, which they may, or may not, allow other men to profit by."[20]

From Education to Ideology

THE CLASSICAL understanding of education is rooted in contemplation and philosophy—i.e., *philo-sophia*, the love of wisdom and the pursuit of truth.[21] St. Thomas Aquinas defined truth as the conforming of the mind to reality and the late Thomist philosopher Joseph Pieper connects truth and prudence.[22] Prudence is seeing reality as it is and making decisions about how to act based on reality: "true humanness—consists in this, that 'reason perfected in the cognition of truth' shall inwardly shape and imprint his volition and action."[23] The classical goal of education that shaped Western civilization was to seek after truth and conform ourselves to it that we could become prudent—and, through prudence, see the world as it is, and be able to exercise the other virtues. Prudence, argued Pieper, was the mother of all virtues because it made them possible. We can only be just, temperate, brave, magnanimous, noble, and show mercy or compassion if we see the world as it is.[24]

The formation of the intellect was intertwined with the formation of the soul and the emotions so that the educated man would not only think clearly, but also live well and have the correct sentiments and emotional responses. Education was seen not merely as passing on knowledge but as training the person to be fully human.

Lewis contrasts this richer understanding of education with the impoverished system of utilitarian and rationalistic education in his fantasy novel *The Voyage of the Dawn Treader*. In our world, Eustace Scrubb and Jill Pole go to a progressive school with the reductionist educational philosophy of the *The Green Book*.[25] Eustace in particular is everything one would expect of such a system. He is scientism's prodigy child, concerned with facts and figures, progress, and engineering, and already though still a schoolboy, a cynic. His imagination is dulled, and he has no knowledge or care for lore, stories, tradition, or ceremony. It is only in Narnia that he at last can benefit from education in the more classical sense and learn that education is more than facts and figures. Education is also about growing in courage, prudence, and sacrifice. From Aslan and the others, Eustace learns what it means to be human.

And yet, for all of Eustace's obvious learning and growth, under the rubric of scientism, none of it counts as learning about anything that actually matters. Since scientism rejects the "reality" of anything non-empirical and rationalism places all things non-empirical outside the realm of reasoned discourse and cognitive access, the entire vision of education is transformed. The stunted Eustace is all this new kind of education can produce.

Under scientism, education loses its "philosophical" dimension. It is no longer oriented to or rooted in truth. In the best-case scenario education becomes simply technical and vocational training where the student is trained in the arts of engineering, plumbing, architecture, medicine and so forth and all other elements of education are handled by families, churches, and other civil associations besides schools and universities. This is not optimal, because moral and aesthetic training should be part of education as a whole, but it would at least be tolerable. As we know, however, this is not what happens. There is no such thing as morally neutral education. Secularism is not a neutral position. What actually happens is that *philo-sophia* is replaced by ideology and education becomes indoctrination.

Education then can no longer can have anything to do with virtue, ethics, citizenry, and freedom. This is most disastrous in the "liberal arts" which are meant to free a man from internal slavery. Liberal arts divorced from truth become power politics and social fashion. One need only look at a university catalog to find the evidence.

If we are going to be prudent and to act well, we have to see well. We have to engage reality. The problem is that scientism rules out entire segments of reality and prevents us from engaging them with our reason.

Plato said that the beginning of philosophy is wonder.[26] St. Thomas Aquinas said that the poet and the philosopher are akin because they both deal with wonder.[27] Lewis argued that modern education often did the very opposite of what it was supposed to—it destroyed wonder. What was needed was a revival of wonder. While many educators of his day saw their role as removing sentiment from the student and getting him to think rationally, Lewis saw a different goal. He argued that education should produce reason in both thought and feeling, and that what was needed was "not to cut down jungles" of emotion, "but to irrigate deserts" to instill correct and reasonable emotions about reality.[28] Education should encourage wonder and proper sentiment. Many students of his day shaped by the likes of the *The Green Book* had their souls and senses dulled by the pedestrian scientism and rationalism of the time. They were like Eustace Scrubb before his Narnian adventure.

Lewis's arguments about the need to revive emotion and sentiment may seem out of step with the problems we face in contemporary culture, where we seem to have nothing but sentiment and everyone talking about their feelings all of the time. Instead of the dry, calculating rationalism that Lewis is addressing, we have dripping sentiment, a celebrity-obsessed culture, and what Theodore Dalrymple has called "toxic cult of sentimentality."[29] People love to talk about narrative, story, passions, and emotions. Few people say what they *think* anymore, it is always about what they *feel*. This emphasis is at least a contributing cause to the fact that Americans score quite low in math and science but very high in self-esteem, with many viewing themselves as quite creative and imaginative. Social media and YouTube are vehicles for this cult of emotional expression. Thus it seems that Lewis was dealing with problems of a different age. Yet on closer examination, one can discern a deep connection.

The emotivism and "toxic sentimentality" of our day is a derivative of the cold scientism and rationalism of Eustace's world. The proverbial saying "nature abhors a vacuum" holds true. If we cannot have reasonable emotions, we will have unreasonable ones. If we do not tutor our sentiments, they don't go away. They become toxic. Emotions become emotivism, sentiment sentimentality, and passions become our master. Lewis understood that "[t]he right defence against false sentiments is to inculcate just sentiments. By starving the sensibility of our pupils we only make them easier prey to the propagandist when he comes."[30] Sentimentality and cold, utilitarian calculation are two sides of the same coin, and one of the goals of education is to tutor the intellect and the emotions to avoid both extremes.

Beauty and the Arts

LEWIS UNDERSTOOD that a scientistic framework not only precludes the possibility of discussing what is just, good, and true, but also severs intelligible contact with that third classic element of human seeking—the beautiful. Beauty is reduced to the "eye of the beholder" and to subjective preference. This has the effect of separating the acts of creating, viewing, and discussing art from the realm of reason and making reasoned criticism of art a futile exercise. If art is no longer intelligible, if it is no longer representative of something outside itself, art is reduced to mere self-expression.

Valuing self-expression may work if you are Rafael or Rembrandt or Donne, but what if you have a boring, banal self? Are you really well served being told your time is best spent expressing yourself as opposed to, say, sitting at the feet of great painters, poets, theologians and philosophers? Similarly, reducing beauty to the "eye of the beholder" seems harmless enough at first glance, but it has consequences beyond aesthetics. De-rationalizing beauty affects education and culture, and though it appears to glorify the self, undermines authentic subjectivity. *The Green Book* begins to lay the foundation of subjectivism not by questioning the morality of stealing or murder, but by addressing questions of beauty. As Lewis notes, real literary and art criticism is difficult, it is much easier to do away with objective beauty altogether. This simple move has profound implications.

Under the limiting rubric of scientism, traditional elements of beauty such as integrity and harmony are not measurable and thus discarded. Only proportion remains, and even that is subject to preference. After all, who is to say what is proportionate? The concept of right proportion may be able to be preserved under some category having to do with evolutionary fitness, but then we aren't really talking about beauty as beauty anymore but merely an impulse for something tenuously connected with physical survival. We can say whether something fits, whether it is long enough or short enough, but the concept of proportionate is relational and includes an element of value judgment and sensibility, and these are outside the scope of reason in scientism. The traditional standards by which we discuss, measure, and critique the beautiful are all reduced to subjective predilection. The idea that students should be taught to appreciate and love the beautiful is removed from education because the feelings, sentiments, and emotions that we experience when moved by a work of art are reduced to mere psychological states disconnected from the work of art.

Now I want to be clear. While I *am* arguing that beauty is not merely in the eye of the beholder and that there are indeed objective standards to measure beauty beyond simply "I like this," I am *not* arguing that ascertaining what is beautiful and why it is so is an easy task. If questions of beauty were simple there wouldn't be a whole branch of philosophy, aesthetics, dedicated to it. But just because beauty—and the appreciation of it—is complicated does not mean that it is merely subjective preference.

I am also not saying that there cannot be innocent reasons that some instances of beauty will resonate with one aesthetically sensitive person more than with another. We may subjectively prefer things that are less beautiful than other things for a host of reasons. We may like a certain work of art that is not particularly beautiful but because our grandfather had it hanging it his living room, or we may like some kitschy song because it reminds us of friends. There are things I like that are not very beautiful but have some other attractive features that resonate with me. I like Georg Telemann's *Trumpet Concerto in D*, which is sublimely beautiful and deeply moving. I also like some country music songs that do not merit the description beautiful. Conversely, there are works of great beauty that we may not appreciate in the least because we lack the tools to engage the work. There are things I found boring as an adolescent that I find deeply moving now. I am told on good authority that the Old English poem Beowulf contains moments of astonishing power and beauty, but since I don't know enough Old English to engage the text as poetry, I can't pick up a copy and enjoy its beauty to any significant degree. My likes tell me what I like; they do not necessarily tell me what is beautiful. As Lewis acknowledged about his own lack of enjoyment of the company of young children: "I recognize this as defect in myself—just as a man may have to recognize that he is tone deaf or colour blind."[31]

It is at times excruciatingly difficult to say why something is beautiful or why one piece or art is more beautiful than another. We may disagree, we may be inconsistent, we may not be able to explain what we perceive. All true. But this is not a reason to reject objective beauty. Answers do not come easy to be sure, but notice that the straitjacket of scientism doesn't even allow the discussion to take place. Scientism creates an impoverished framework so that those subjected to its narrow horizons lose the capacity required to perceive the diverse complexity of reality. They become blind as it were to anything that does not fit within their pre-conceived intellectual framework and thus are unable to intellectually process whole swaths of reality. Lewis represents this inability to see a larger, more complex reality in his novel *Out of the Silent Planet*, when Devon and Weston are unable to process and comprehend the words and meaning of the Oyarsa of Malacandra because he does not fit into their scientistic framework.[32] In the indiscriminate hands of the proponent of scientism, Ockham's razor becomes a hacksaw that instead of clarifying severs the mind from reality.

Destruction of Authentic Subjectivity

THE LOSS of objective beauty that flows from scientism has an impoverishing effect on culture, but even more insidious is its effect on how we understand human subjectivity.

When beauty is reduced to the eye of the beholder, it deifies the beholder. Individual opinions become paramount. What matters is not some tradition or objective standard, but the unique subject and his own feelings and emotions. But it's really a devil's bargain since scientism, by destroying the possibility of reasonable and objective beauty, actually undermines the value of subjective experience. When beauty is reduced to the merely subjective, we take a profound truth—that each person is unique and unrepeatable and so experiences a beautiful thing uniquely—and we reduce it to the banal observation that everybody is entitled to his own opinion, failing to realize that in emptying beauty of anything objective, scientism leaves the beholder with nothing but his opinion. When beauty is rational and objective, subjective experience matters because it brings a perspective about something "out there" which can be communicated to others. Perhaps one person sees something that the other didn't see, or perhaps he misunderstood the piece and another can explain it to him. When different people see different aspects and are able to communicate to one another in a reasonable way the souls of both can be expanded. But this is all impossible—or at least non-rational and silly—under the auspices of scientism. If the emotional reaction to beauty is not reasonable and intelligible, if we are not actually speaking about something, then what we feel about that piece of art or music is absolutely unimportant, and the whole idea of communicating to others makes no sense.

If all you are doing is talking about your feelings completely disconnected from the outside world, then unless I have some ulterior motive for listening—e.g. you are my boss, I want you to tweet my new idea on art criticism—then I will have no motivation to care other than maybe to sympathize with you if your feelings are those of excruciating pain.

So this is the choice: Under a broad, classical understanding of reason, if you tell me your feelings about a piece of music, what it means to you and why you think it's beautiful, ugly, sublime, pretty, or boring, I can agree or disagree with you, I can think you are dead wrong, but at least I can take you seriously and we can have a real discussion. But when emotions are no longer viewed as rational, it undermines subjectivity and communication. It economizes relationships. In order for me to be interested in what you have to say I have to have some utilitarian motive since you aren't really saying anything reasonable.

One could protest that scientism doesn't really have these effects, and this is true to some degree, but that is only because no one actually lives his life fully aligned with scientism's truncated view of reason, just as no one actually lives like a strict relativist. But scientism is still a problem. When our intellectual framework does not match our lived experience, it creates a divided man. Instead of conforming our theory to reality we try to shrink reality to fit into our hobbled theory, which in this case ultimately leads to irrationalism when it comes to value judgments and emotional life. And an irrational emotional life vitiates authentic subjectivity. Feelings, passions, and emotions are reduced to hormones, chemical reactions, the result of genetics or the social structures that form us. Because passions, feelings, and emotions are not empirical and not rational, concepts like subjectivity personhood, personality, and temperament must be explained by something physical, or they must be discarded. Thus what appears to be the glorification of subjectivity and the self actually abolishes them. The poison of subjectivism leads to the death of the subjective. Once we accept a truncated vision of reason, it gives us a narrow vision of life. Simply put, reason matters.

MEN WITH CHESTS: REBUILDING THE MORAL IMAGINATION

THE QUESTION THEN IS HOW to we rehabilitate reason and all the things that go along with it. Lewis spent a much of his intellectual energy making clear, reasoned arguments, and helping people to think clearly. He realized that a lot of the emotional, psychological, and social problems of his day were the result of muddled thinking. Yet he also wrote fantasies and fairy tales because he realized that intellectual arguments, though essential, are insufficient in themselves. Even for a coherent thinker, clear, reasoned argument is not sufficient for spurring a man on to do the right thing in the face of tremendous pressure. As Lewis put it: "In battle it is not syllogisms that will keep the reluctant nerves and muscles to their post in the third hour of the bombardment."[33]

A human being is more than just an intellect, and a muddled intellect usually means muddled emotions, morals, and spiritual life. A clear, reasoned argument will not penetrate a mind muddled by disordered emotional and moral attachments. Many of our intellectual problems with truth and reason are not intellectual problems at all—they are moral and spiritual problems that have gone looking for a rationalization to justify them. Staked out intellectual positions are not only activities of our intellect, but tied in to our emotional and spiritual lives. As Lewis argued, the "head rules the belly through the chest," the seat of emotion and sentiment, or what is often called the heart. In other words, the intellect can only tame and tutor the appetites, the will, if it first enlists the chest. In order to rehabilitate reason, we need to make strong arguments, but we also have to rehabilitate the imagination and the heart.

This is not an easy or a simple task, but I will briefly discuss several things we can do to begin to form the moral imagination anew.

1. Read Good Stories

THE FIRST thing that is required to re-build the moral imagination and prepare the heart for "reasonable emotions" is to read and hear good stories. This is essential for children, but it is not only for children. Adults too need to purify and renew their imaginations. Lewis's own work was a mix of solid arguments and good stories, with the stories helping prepare the mind to understand the arguments. Fairy tales resonate with children because they naturally see the world in terms of good and evil. One of the main questions my six-year-old daughter asks me when a new character appears is "Is he good or bad?" Children have not yet been subjected to relativism and the likes of *The Green Book*, and good stories help develop the moral imagination and prepare the child to embrace clear thinking.

Several fine books have been written about the importance of fairy tales and their role in shaping the moral imagination. Vigen Guroian wrote a short book called *Tending the Heart of Virtue*, and Mitchell Kalpakgian wrote another fine book called The *Mysteries of Life in Children's Literature*.[34] The late John Senior argued that before reading the great books we should read "good books." Great books, he said, were important, but perhaps even more important to help develop our humanity and moral imagination were what he called the good books.[35]

All of these writers recognize that developing the moral imagination cannot be reduced to simply reading good stories, but good stories are essential, especially for children who are being formed. Why? What do good stories (especially fairy stories) do? Fairy tales and other good stories set out a moral universe, and they teach truth about reality. Aquinas says truth is conforming the mind to reality, and fairy tales are often the best of stories at showing that good is good, evil is evil, and that our actions have consequences.

Guroian and Kalpagian discuss the difference between good and bad wishes. Good wishes are rooted in justice and desire for truth and goodness; bad wishes are rooted in pride and greed. Cinderella, despite being subjected to injustice and torment, remains virtuous and prays at the grave of her mother for divine intervention. Her wish is granted on the condition she obeys, which she does, though with some difficulty. Her wish is granted; she is able to marry the prince because of her virtue. Bad wishes rooted in pride and greed on the other hand bring destruction and make you lose what you had. In "The Fisherman's Wife," the title character begins with a simple wish for a better place to live, but instead of being content, she grows greedy and covetous, wanting next a mansion, then a castle and then to be queen and pope and finally the enchanted fish has had enough and returns her to her leaky hovel. Fairy tales fables, parables and classical myths also warn about the danger of getting what one wishes for. King Midas's wish for everything he touches to become gold is granted and in his avarice he loses the beloved daughter. Actions have consequences and our desires and emotional attachments matter. Guroian writes: "A person's decision in life defines what kinds of person she becomes."[36]

"Beauty and the Beast," especially the French version by Madame Beaumont[37] is beautifully told. It is reflective of the love of Christ and the powerful myth of Eros and Psyche that Lewis retells in *Till We Have Faces*.[38] What is conveyed in "Beauty and the Beast"? There are three sisters. They are all very beautiful, but the youngest, Beauty, is most beautiful of all. She is also the most virtuous. She thinks of others, even her wicked sisters, sacrifices herself for her father, and through her virtue is able to see beyond the physical surface to the moral goodness of Beast, unlike her sisters who married for money and wit. Her virtue and vision are rewarded, and she marries happily while her envious sisters are turned into stone and forced to observe her happiness.[39] Among the many lessons one learns is that life is much more complex than the physical appearance and that interior virtue is superior to external appearance.

Fairy tales are imbued with an expansive concept of reason and goodness. They are not limited by reductionist scientism but instead present a richer and more complete picture of reality. To those who would dismiss fairy tales as unrealistic for their "happily ever after" endings, I would offer two points. One, anyone offering such an objection hasn't read many classical fairy tales. Many of them, as already noted, don't end happily at all. Second, while the virtuous do not always get the prince and the beautiful castle in this life, fairy tales distills the cosmic truth that goodness wins in the end. In this way fairy tales teach reality in a far more expansive way than, for instances, the stories of nihilism and despair typical of so much naturalistic and modern literature.[40]

Fairy tales have this tutoring capacity for both adults and children. Good stories affect adults too. As Lewis and J. R. R. Tolkien both argued, it is childish to think that good fairy stories are only for children.[41] Just recently I was reading George MacDonald's *Princess and Curdie* and could feel something awaken in me from my childhood—a vision to be better, to strive for moral excellence and pay closer attention to people and the details of reality—to lead a more integrated life. Good stories are essential because they form moral imagination and, equally important, prepare us to recognize the difference between "toxic sentimentality" and reasonable emotions.

2. Detox: Get the Junk Out and Re-sensitize Ourselves to Good and Evil

WHILE READING good stories is quite an enjoyable way to build the moral imagination, the second activity requires more virtue and effort. Reading good stories is one side of the coin. The other is detoxifying our imaginations and re-sensitizing ourselves to good and evil. It is inescapable that modern media, music, and film portray a reductionist vision of the world dominated by scientism and relativism. Love is reduced to sex and/or emotion. Freedom is merely exercising your will, often against authority, whether legitimate or illegitimate. Charity and kindness are reduced to amoral tolerance. A combination of utility, pleasure, and equality is the highest value. Authenticity means following whatever whim you have as long as it conforms to whatever is socially acceptable.

A couple of months ago I did a little experiment on the plane as I was flying from New York to San Francisco to give a talk on the moral imagination. I watched contemporary television shows. I don't watch that much television, but since I was working on a lecture on education and the moral imagination, I decided to see what kinds of things were shaping the imagination of young and old. I chose three prime-time television shows, one called *Suburbgatory*, another called *The Big Bang Theory*, and the third, of which I had heard, *Glee*. These are the types of material that form and shape the moral imagination of millions of TV watching Americans. I assume there are worse things on, but what I saw was unbelievable. Here on prime-time were explicit themes of sensuality, masturbation, a type of dualist anthropology where the body is distinct from your "person," gratification as the highest value, homosexuality equated with creativity, controlling one's body interpreted as having sex whenever and however you want it, relativism, Darwinism, vulgarity, objectification of women, and the high virtues of egalitarianism and tolerance all covered in banal sentimentality with happy music to make you feel good about yourself.

If we're taking this stuff in on a regular basis, then all of this shapes our moral imagination and our sensibilities, and in an indirect way so that we are not aware of what is happening. We get desensitized to evil. We begin to see things that are morally and socially destructive as normal. Lewis said the way we inoculate people against false patriotism and radical nationalism is by training them in true patriotism. People are trained to recognize counterfeit bills by studying authentic ones. But what most of modern television, music, and film gives us is counterfeit emotions instead of real ones, counterfeit sexuality instead of real love, all wrapped up in the worldview of scientism and relativism. One of the reasons Lewis and Tolkien wrote so much fantasy is they recognized the dearth of good modern literature and so decided to write their own.

Just as good stories prepare the mind and heart to accept the classical understanding of reason and reasonable emotions, so banal television prepares the mind and heart to accept the hobbled view offered by scientism. It is not enough just to read good stories. It's not enough just to have strong moral arguments. We also have to stop watching and listening to things that corrupt our imaginations. One might object that this is overly prudish, and we need to engage the culture. But we can only engage a culture when we are not formed by it. The good thing is that when we read beautiful stories and listen to beautiful music, and take the time and energy to wean ourselves from the rubbish, all of a sudden banal television shows aren't as interesting anymore.

3. Recover Objective Beauty

THE RECOVERY of objective beauty is both a philosophical point and exercise in practical training of our aesthetic tastes—and both are difficult. This is the recovery of objective beauty. As I have already discussed, our deep-seated aesthetic subjectivism is an enormous obstacle to building up our moral imagination and to education. As Stephen Faulkner has argued, "If the definitions of 'good' dance and music 'give us the slip and get away,' Plato said, 'it will be pointless utterly to prolong our discussion of correct education, Greek or foreign.' If Plato is right, educational reform in this country is in a lot of trouble."[42] It is essential that we recover an objective sense of the beautiful and that we present beautiful architecture, music, literature, art, and poetry to ourselves, our students, and to our children, so that these things will impress themselves upon our characters and develop our sensibilities.

Another reason the recovery of objective beauty is so essential is that subjective beauty is so intimately connected with scientism and relativism. In *The Abolition of Man*, when Lewis retells the story of the poet Coleridge at the waterfall, it is the argument for a subjective view of beauty that opens the door into subjectivism and relativism. But it can also open the door back. One of the successful strategies I found when teaching undergraduates was to challenge scientism and relativism first by asking them to go back to their own experience. Does this relativistic or naturalist intellectual construct that you have been taught—or what is more accurate, what you have absorbed unwittingly over time—correspond with how you experience the world, how you experience your own emotional responses? Before we engaged questions of ethics, I would begin with listening to music and ask them to pay attention to their emotional experiences, as I would play portions of different pieces from different genres. As we would discuss their experiences it became clear to them that their emotions were responses to different things; that the music had some inherent value or quality that they were reacting to. In time the actual reality of the emotional experiences began to contradict the subjectivism they had been taught and the wall of subjectivism began to tumble—through aesthetics; and further through discussions of love. When it became clear that love and beauty were not merely subjective, subjectivity in morality became a lot less plausible.

In classical and Christian philosophy, beauty plays an important role. We neglect it at our peril. Medieval theologians viewed the good, the true, and the beautiful as essential elements of being—ontologically the same as being but conceptually distinct. Experience of the beauty has the possibility of opening up our souls to the transcendent. Beauty was considered a pure perfection, like love and truth and goodness, and a way to God. As Hans Urs von Balthazar wrote, "Every experience of beauty points to infinity."[43] But this way to God is blocked in a rationalist framework. Re-capturing objective beauty plays an important part in forming the moral imagination and rehabilitating reason.

4. Re-build Authentic Subjectivity

As I discussed earlier, objective beauty is closely related—at least in our experience—to the recovery of authentic subjectivity. This is not easy, but because we live in a highly emotive and sentimental culture, we actually have an advantage here, especially with adolescents and young adults. There is nothing college and high schools students want more than to be taken seriously. And there are few things more important to them than their emotions. And if we can show them that without objective reality their feelings and emotions are unimportant, they become open to argument for objective reality. Subjectivity is intertwined with their understanding of personhood and morality so it has broad personal and social impact.

5. Rehabilitate the Heart

INTIMATELY CONNECTED with the recovery of authentic subjectivity is the recovery of what Lewis calls "reasonable" emotions. Similarly, Dietrich von Hildebrand referred to a type of "intelligible" affectivity.[44] There can tend to be a false dichotomy even among Christian thinkers between the passions and reason. This is too complex a topic to address in this essay, but Lewis touches on this in *The Abolition of Man*, when he discusses the concept of reasonable emotions. He argues that intellect alone is often insufficient to inspire us to moral greatness or simply to fulfill our duty. "Reason in man must rule the mere appetites by means of [what Plato calls] 'the spirited element,'" by "emotions organized by trained habit into stable sentiments… [that] are the indispensible liaison between cerebral man and visceral man."[45] These are emotions that are distinct from raw passion. As Von Hildebrand describes it, we respond emotionally to the intrinsic value in a person, and act of heroism or mercy, or a beautiful piece of art. The affective response happens with "a collaboration of our intellect"; it is "not mere sense perception but implies a full actualization of our intellect"[46] and we say "yes" to the emotion—or "no" if it is unreasonable.[47] It does not "overwhelm" us; rather we participate in it. We can "decapitate" or "sanction" the affective response.[48] It is a reasonable emotional response to the good, the true, and the beautiful. These emotions are not mere appetites or whims; they are deep emotional responses connected to reality and to reason.

It is through these reasonable emotions that one can integrate the passions and order them according to reason. As Lewis wrote: "The head rules the belly through the chest."[49] It is the chest that gives us a vision of moral greatness, inspires us to nobility, heroism, compassion, mercy, justice, and love. It is the chest, the reasonable emotions, which are moved when we read fairy tales and see acts of sacrifice. And when all of our fears or desires may be pushing us toward selfishness, cowardice or gratification, it is the chest, "furnished from the wardrobe of a moral imagination, which the heart owns and the understanding ratifies as necessary to cover the defects of our naked, shivering nature and to raise it to dignity"[50] that enables us to rise above our appetites and lead a noble, authentic human existence. This is the classic vision of the good life—a life of virtue in accordance with reason where we take pleasure in doing the good. Right now we have big bellies and intellects hobbled by scientism. One of our tasks is to renew the moral imagination and rebuild the chest. In so doing, our lived experience of reasonable emotions becomes an intellectual and experiential refutation of truncated scientism. It is here that we especially see the interconnectedness of clear arguments, good stories, and the renewal of the moral imagination.

6. Silence

THERE ARE several other things that play an important role in rebuilding the moral imagination. Re-orienting our worship and renewing our appreciation of the sacred is one, and recovering language and the connection between words and reality is another. These would take too long to develop in this chapter, but I think they are worth mentioning.[51] I do, however, want to briefly address one final element for renewing the moral imagination and which plays an important, though perhaps underappreciated, role. That is silence.[52] Silence is always important for mental and spiritual health, but it is even more crucial our current culture that we cultivate silence. There is something about the modern, highly technical way we experience life that makes it easy for us to be inundated with information and constant communication. Far from broadening us, this can contribute to a narrow and compartmentalized vision of the world. This is not necessarily the case, but it can encourage an instrumental understanding of reason. Lewis and Tolkien were known for taking long walks and for paying close attention to nature and to people, which is one element that made their descriptive fiction so rich.

Cultivating silence enables us to be receptive to elements of reality that are not instrumental or may not have a utility but may in fact be imbued with meaning. Because silence can enable us to be attentive to certain complexities both outside of us and within—things that are real and reasonable but which cannot be explained by empirical scientism—it can open our horizons and prepare the intellect and heart to be receptive to the expansive classical and Christian understanding of reason.

My goal in this last section was only to introduce some of the things we can do to rebuild the moral imagination and help prepare the mind and heart to be ready to receive and understand the "doctrine of objective values" and a rich, expansive vision of human reason. Teaching someone in college and graduate school why scientism and relativism are incoherent becomes much easier when he or she has been shaped by good stories, beautiful art, and silence, instead of *The Green Book* and television sitcoms.

Conclusions

The task that C. S. Lewis put before us—to rehabilitate reason—is not a simple one. It will require a multi-faceted approach to escape from the straightjacket of reductionism. Part of our challenge in a reductionist milieu is to not promote single solutions to the problem of scientism. Sound argument and the formation of the emotions, the virtues, the moral imagination, and the soul are all essential.

It is popular today to argue for the importance of story and narrative to convey truth. Yes, but narrative unhinged from truth and reason can be equally used to convey a lie. Dostoyevsky famously wrote that "beauty would save the world."[53] But not beauty severed from reason. Strong emotional appeals are necessary, but unless they are grounded in reasonable emotions they can easily become "toxic sentimentality." Clear and forceful intellectual argument is essential but will not pass through ears and an intellect unprepared to hear. God's grace can of course overcome any obstacle, but grace also builds on nature and the seeds of faith tend to grow best when it lands on an intellect, soul and heart disposed to hear it.

Finally, faith is necessary to rehabilitate and "purify" reason. As Pope Benedict said in his Regensburg Address on the crisis of reason in the West: "We will succeed in doing so only if reason and faith come together in a new way, if we overcome the self-imposed limitation of reason to the empirically falsifiable, and if we once more disclose its vast horizons."[54]

Endnotes

1. C. S. Lewis, *The Abolition of Man* (New York: Macmillan, 1955), 13.
2. Ibid., 16–17.
3. Ibid., 34.
4. Ibid., 35.
5. Edmund Burke, *Reflections on the Revolution in France* edited by J. G. A. Pocock (Indianapolis: Hackett Publishing, 1987), 67.
6. See G. K. Chesterton, *Orthodoxy* (Hollywood, FL: Simon and Brown, 2012)
7. Lewis, *Abolition of Man*, 30.
8. Christopher Dawson, *The Historic Reality of Christian Culture* (New York: Elliots Books, 1960), 90.
9. Lewis, *Abolition of Man*, 16–17.
10. Mark Roberts, "Facts, Opinions and Values: Philosophical Distinctions and Reading Skills," paper read at the conference of the Society of Catholic Social Scientists, Franciscan University of Steubenville, October 1997.
11. Lewis, *Abolition of Man*, 13.
12. Much of the following draws from Roberts's presentation, which I attended.
13. Lewis, *Abolition of Man*, 16.
14. Ibid., 22.
15. Ibid., 25.
16. Ibid., 20.
17. Joseph Ratzinger, "Political Visions and Political Praxis," in *Values in a Time of Upheaval* (San Francisco: Ignatius Press, 2006), 24.
18. Ibid.
19. Lewis writes in his preface to *That Hideous Strength*: "This is a 'tall story' about devilry, though it has behind it a serious 'point' which I have tried to make in my *Abolition of Man*." Lewis, *That Hideous Strength* (New York: Macmillan, 1965), 7.
20. Lewis, *Abolition of Man*, 68. See also: "Man's power over Nature turns out to be a power exercised by some men over other men with Nature as its instrument." Ibid., 69.
21. See Josef Pieper, *Leisure, the Basis of Culture* (South Bend, IN, St. Augustine's Press, 1998).
22. See Josef Pieper's *The Four Cardinal Virtues* (Notre Dame, IN: University of Notre Dame Press, 1966), especially 3–40.
23. Ibid., 8.
24. See ibid., chapter 2.
25. C. S. Lewis, *The Voyage of the Dawn Treader* (New York: Macmillan, 1952).
26. Plato, *Theaetetus*, Seth Benardete, trans. (Chicago: University of Chicago Press, 1984), I.19, section 155d.

27. George William Rutler, "Poets and Philosophers Are Alike in Being Big with Wonder," Catholic Education Resource Center, accessed June 27, 2012, http://www.catholiceducation.org/articles/catholic_stories/cs0428.htm.
28. Lewis, *Abolition of Man*, 24.
29. Theodore Dalrymple, *Spoilt Rotten: The Toxic Cult of Sentimentality* (London: Gibson Square Books, 2011).
30. Lewis, *Abolition of Man*, 24.
31. Ibid., 29.
32. C. S. Lewis, *Out of the Silent Planet* (New York: Macmillan, 1965), 125–132.
33. Lewis, *Abolition of Man*, 34.
34. Vigen Guroian, *Tending the Heart of Virtue* (Oxford: Oxford University Press, 1998); Mitchell Kalpagian, *The Mysteries of Life in Children's Literature* (Long Prarie, MN: Neumann Press, 2000). Also see Andrew Pudewa's talks on the moral imagination at the Institute for Excellence in Writing, www.excellenceinwriting.com.
35. John Senior developed a list of what he called "The 1000 good books" from pre-school on. See John Senior, *The Death of Christian Culture* (Norfolk, VA: IHS Press, 2008) and *The Restoration of Christian Culture* (Norfolk, VA: IHS Press, 2008). The list of the 1000 is in the appendix of *The Death of Christian Culture*.
36. Guroian, *Tending the Heart of Virtue*, 22.
37. Alfred and Mary Elizabeth David, eds., *The Twelve Dancing Princesses and Other Fairy Tales* (New York: New American Library, Signet Classic, 1964).
38. C. S. Lewis, *Till We Have Faces* (New York: Harvest/HBJ, 1980).
39. Lewis, *Till We Have Faces*; Guroian, *Tending the Heart of Virtue*, 21–23; and Kalpagian, *Mysteries of Life in Children's Literature*.
40. Thanks to Jonathan Witt for suggesting this additional point.
41. See Tolkien's "On Fairy Stories," *Tree and Leaf* (London: Harper Collins, 2001). See also Lewis's short essay, "Sometimes fairy stories may best say what's to be said" (1956), accessed June 5, 2012, http://wedgwoodcircle.com/news/articles/sometimes-fairy-stories-may-say-best-whats-to-be-said/.
42. Stephen Faulkner, "The Workshop of Worship," *Touchstone*, 9, no. 2 (Spring 1996), accessed June 25, 2012, http://touchstonemag.com/archives/article.php?id=09-02-031-f.
43. Quoted in Thomas Dubay, *The Evidential Power of Beauty: Science and Theology Meet* (San Francisco: Ignatius Press, 1999), 3.
44. See Dietrich von Hildebrand, *The Heart* (South Bend, IN: St. Augustine Press, 2007) and *The Nature of Love* (South Bend, IN: St. Augustine's Press, 2009). Dietrich and Alice von Hildebrand, *The Art of Living*, (Chicago, Franciscan Herald Press, 1965) and *Transformation in Christ*, (San Francisco, Ignatius Press, 1998). I am grateful to Dr. Michael Healy and Dr. John F. Crosby for many of these insights.
45. Lewis, *Abolition of Man*, 34.
46. Dietrich and Alice von Hildebrand, *The Art of Living* (Chicago, Franciscan Herald Press 1965), 108.
47. Ibid., 117.

48. Dietrich von Hildebrand, *Transformation in Christ* (San Francisco: Ignatius Press, 2001), 225.
49. Lewis, *Abolition of Man*, 34.
50. Burke, *Reflections on the Revolution in France*, 67.
51. One very short book I recommend is Josef Pieper's *Abuse of Language, Abuse of Power* (San Francisco: Ignatius Press, 1992).
52. I am grateful to Sr. Thomas More Stepnowski OP for her comments and suggestion to include a discussion of silence. For more discussion of this topic, see Max Picard, *The World of Silence*, second edition (Eighth Day Press, 2002).
53. Fyodor Dostoevsky, *The Idiot*, translated by Eva Martin, Part II, Chapter V, accessed June 25, 2012, http://en.wikisource.org/wiki/The_Idiot/Part_III/Chapter_V.
54. Pope Benedict XVI, "Faith, Reason and the University: Memories and Reflections, Meeting with Representatives of Science," Regensburg, Germany, September 12, 2006, accessed June 5, 2012, http://www.vatican.va/holy_father/benedict_xvi/speeches/2006/september/documents/hf_ben-xvi_spe_20060912_university-regensburg_en.html.

Biographies of Contributors

John G. West

John G. West co-edited the award-winning *C. S. Lewis Readers' Encyclopedia* (Zondervan) and is author of several other books, including *Darwin Day in America* and *The Politics of Revelation and Reason*. Dr. West is a Senior Fellow at the Seattle-based Discovery Institute, where he is Associate Director of the Center for Science and Culture and Director of the C. S. Lewis Fellows Program on Science and Society. Formerly the Chair of the Department of Political Science and Geography at Seattle Pacific University, Dr. West holds a Ph.D. in Government from Claremont Graduate University.

M. D. Aeschliman

M. D. Aeschliman is Professor of Education at Boston University, Adjunct Professor of Anglophone Culture at the University of Italian Switzerland (Lugano), and the author of *The Restitution of Man: C. S. Lewis and the Case Against Scientism*. He is a widely-published scholar and literary critic whose articles can be found in the pages of *First Things*, *National Review*, and many other publications. Dr. Aeschliman holds a Ph.D. from Columbia University.

Jake Akins

Jake Akins is co-editor with William Dembski of a forthcoming anthology on natural theology during the British Enlightenment. He holds a degree in political science and has done graduate work in theology. He resides in Texas.

C. John Collins

C. John Collins is Professor of Old Testament at Covenant Theological Seminary in St. Louis and author of *Science and Faith: Friends or Foes?*,

The God of Miracles, and *Did Adam and Eve Really Exist?* among other books. Dr. Collins teaches a seminary course about the life and writings of C. S. Lewis, and he is also a Fellow with the Center for Science and Culture at Discovery Institute. Dr. Collins holds a Ph.D. from the University of Liverpool and a M.S. from MIT.

JAMES A. HERRICK

James A. Herrick is the Guy Vander Jagt Professor of Communication at Hope College in Holland, Michigan. He is author of such books as *The Making of the New Spirituality, Scientific Mythologies,* and *The Radical Rhetoric of the English Deists.* He writes and speaks about spiritual themes in popular culture, the history of skepticism, new religious movements, and popular narratives about science and progress. Dr. Herrick is a founding member of the Baylor University Press series in rhetoric and religion. He holds a Ph.D. from the University of Wisconsin.

PHILLIP JOHNSON

Phillip E. Johnson is the Jefferson E. Peyser Professor of Law (Emeritus) at the University of California, Berkeley School of Law and the author of numerous books, including *Darwin on Trial, The Right Questions,* and *Reason in the Balance.*

EDWARD J. LARSON

Edward J. Larson is Hugh and Hazel Darling Chair in Law at Pepperdine University School of Law. He is author of numerous books, including *An Empire of Ice: Scott, Shackleton, and the Heroic Age of Antarctic Science; A Magnificent Catastrophe: The Tumultuous Election of 1800;* and the Pulitzer Prize-winning *Summer for the Gods: The Scopes Trial and America's Continuing Debate Over Science and Religion.* Dr. Larson holds a Ph.D. from the University of Wisconsin and a J.D. from Harvard University.

MICHAEL M. MILLER

Michael M. Miller is a Research Fellow and Director of Acton Media at the Acton Institute. He previously chaired the philosophy and theology department at Ave Maria College of the Americas in Nicaragua, where he also taught courses in philosophy and political science. He holds graduate degrees in philosophy, international development, and international business.

VICTOR REPPERT

Victor Reppert is author of *C. S. Lewis's Dangerous Idea*, the definitive book about Lewis's argument from reason, which has been praised by noted author Peter Kreeft as "philosophical revisionism at its finest." Dr. Reppert teaches at Glendale Community College in Arizona and has written numerous articles examining the philosophical arguments of C. S. Lewis. Dr. Reppert holds a Ph.D. in Philosophy from the University of Illinois.

JAY RICHARDS

Jay Richards is co-author of the *New York Times*-bestseller *Indivisible* as well as *The Privileged Planet: How Our Cosmos is Designed for Discovery*, which inspired an acclaimed science documentary of the same name. He is also editor of the award-winning book *God and Evolution: Protestants, Catholics, and Jews Explore Darwin's Challenge to Faith*. Dr. Richards is a Senior Fellow with Discovery Institute, where he directs the Center on Wealth, Poverty, and Morality and is a Senior Fellows with the Center for Science and Culture. Dr. Richards holds a Ph.D. from Princeton Theological Seminary.

CAMERON WYBROW

Cameron Wybrow has taught Religion, Politics, Classical Civilization, Great Books, Greek, and Hebrew at various universities. He is the author of two scholarly books on Christian creation doctrine and the rise of modern natural science, and of a number of scholarly articles in political philosophy, Biblical studies, and intellectual history, as well as many book reviews and popular articles on religion, education, and intelligent design. Dr. Wybrow holds a Ph.D. in Western Religious Thought from McMaster University.

INDEX

A
Abolition of Man, The 9, 19, 39, 47, 63, 75, 144, 180, 213, 238, 239, 240, 244, 252, 257, 258, 265, 293, 309, 311, 318, 331, 333
Absolute idealism 203
Acworth, Bernard 113, 137
Adam and Eve 121
Anscombe, Elizabeth 183, 199, 211, 219
  disagreements with Lewis 212
Applegate, Kathryn 166
Aquinas, St. Thomas 49, 189, 295, 318, 320
Argument from computers 226
Argument from reason 199, 203, 204, 210
  as the best explanation 214
Aristotle 80, 295
Astrology 9, 61, 63
Augustine 268
Autobiography (Charles Darwin) 11, 111, 130
Ayer, A. J. 295, 303

B
Bacon, Francis 49
Bailey, Ronald 254
Balfour, Arthur J. 129, 188, 211
Balthazar, Hans Urs von 332
Barefoot, Darek 225
Barfield, Owen 48, 64, 203
Barlow, Connie 22
Beauty 322, 331
Behe, Michael 158, 159, 161, 172
Bergson, Henri 124, 134
Bernal, J. D. 238, 247
Beversluis, John 209, 224
Beyond Freedom and Dignity 302

BioLogos Foundation 70, 110, 115, 157, 166
Blackmore, Susan 207
Bostrom, Nick 252, 253
Brave New World 272
Bryan, William Jennings 109
Bundy, Ted 206
Burke, Edmund 310

C
Calabria, Giovanni 122
Carritt, E. F. 69
Cashmore, Anthony 94
Chance or Design? 159
Chardin, Pierre Teilhard de 135, 251
Chesterton, G. K. 120, 128, 246, 311
Chesworth, Amanda 22
Christian Theology and Natural Science 224
Churchland, Patricia 190
Clarke, Arthur C. 242, 246
Colet, John 87
Collins, C. John 171
Collins, Francis 70, 110, 139, 143
Conditioners 241, 271, 317
Control of Language, The (aka The Green Book) 267
Copernican Revolution 64
Copernicus 72
Coyne, George 133
C. S. Lewis Company 13
C. S. Lewis's Dangerous Idea 199, 213, 224

D
Dalrymple, Theodore 320
Darrow, Clarence 109
Darwin, Charles 9, 11, 74, 109, 111, 119, 123, 128, 163, 186, 210
Darwin Day 22

Darwin on Trial 144
Darwin's Black Box 158
Darwin's doubt 190
Davos Man 317
Dawkins, Richard 10, 21, 24, 123, 139, 162
Dawson, Christopher 311
Deep Blue computer 227
Deep Ecology 48
Dembski, William 158, 159, 161, 172
Dennett, Daniel 205
Descartes 210
Descent of Man, The 119, 128
Design argument
  functional complexity 156
  morality 155
  natural beauty 154
  reason 155
Determinism 201
Discarded Image, The 11, 53, 55, 59, 60, 65, 71, 73, 141
Discovery Institute 161
Dostoevsky, Fyodor 295, 335
Dowd, Michael 22
Draper, Paul 221
Duhem, Pierre 49
Dymer 11, 31, 48
Dyson, Freeman 255
Dyson, Hugo 96

E

Eco-feminism 48
Eliot, T. S. 294
Emergent Self, The 224
"Empty Universe, The" 55
English Literature in the Sixteenth Century 54
Eugenics
  Nazi 28
  Shaw's views 246
Everlasting Man, The 120, 128
Evolution
  as common descent 113
  as myth 20, 77, 92, 136
  as natural selection 122
Evolutionary argument against naturalism (EAAN) 190
Evolution: A Theory in Crisis 159
Evolutionism 26, 77, 136
Evolution Protest Movement 137

F

Fall of Man 90, 115, 236
Farrer, Austin 97
Finding Darwin's God 110, 133
First Cause argument 188
First Things 13
Flew, Antony 172
Ford, Paul 129
Formby, C. W. 116
Foundations of Belief, The 129, 211
Freedom of the Will 224
Freudianism 26
Freud, Sigmund 9, 11, 25, 74, 209
Fukuyama, Francis 257
"Funeral of a Great Myth, The" 54
Futuyma, Douglas 139

G

Galileo 59, 64, 72
Galton, Francis 27, 238
Gardner, Helen 121
Garis, Hugo de 253, 255
Genesis of Species, The 128
Giberson, Karl 110, 115
Gilder, George 40
God of the gaps argument 228
Gribbin, John 86
Guroian, Vigen 327

H

Haecker, Theodor 293
Haldane, J. B. S. 30, 33, 71, 76, 79, 185, 190, 201, 238, 242, 243, 248
Harris, John 254
Harwood, A. C. 48
Hasker, William 215, 224
Hildebrand, Dietrich von 332
Holderlin, Friedrich 294
Hughes, James 252
Human enhancement 235, 255
Huntington, Samuel 317
Huxley, Aldous 272

Huxley, Julian 28, 30, 34
Huxley, Thomas H. 47, 245
hyper-Freudianism 209

I
Inherit the Wind 109
Intelligent Design 158
Intelligent Design 40, 83, 110, 123, 131, 134, 137, 143, 159, 167, 172
  Lewis's views 9, 82, 156, 157, 165
"Is Progress Possible?" 57

J
Jaffa, Henry 302
Jaki, Stanley L. 49
James, William 297
Johnson, Phillip 23, 144
Jonas, Hans 299
Jones, Jr, Bob 70

K
Kalpakgian, Mitchell 327
Kant, Immanuel 210
Keller, Tim 121
Ketley, Martin 267
Keynes, John Maynard 304
Kierkegaard, Soren 295
King, Alec 267
Kinsey, Alfred 28
Kurzweil, Ray 254

L
Language of God, The 70, 110
Laplace 228
Larmer, Robert 228
Leibniz, Gottfried Wilhelm 297
L'Evolution Creatice 125
Lewontin, Richard 139
Lieberman, Daniel 37
Lovelock, James 37
Lucas, J. R. 224
Lyell, Charles 163

M
MacDonald, George 329
Magic 9, 19, 24, 29, 38, 48, 57, 61, 135, 166, 227, 242, 253, 260
Magician's Nephew, The 19, 38

Man's conquest of nature 271
Maritain, Jacques 294
Marx, Karl 9
Mascall, Eric 224
Materialism 200, 201, 204
  defenses of 206
  existence purposeless 205
  makes universe empty 204
  no norms of rationality 206
  no norms or morals 206
  No place for subjective 205
Matter and Spirit 211
McLaren, Brian 120
Menuge, Angus 225
Men without chests 240, 269, 273, 310, 311, 313, 314, 326
Mere Christianity 84, 111, 134
Meyer, Stephen 158, 163
Middle Ages 50, 59
Miller, Kenneth 110, 133
Miracles 55, 91, 120, 129, 132, 180, 184, 201, 211, 213
Mivart, George Jackson 128
Moral imagination 309, 310, 311, 326, 329, 331
Moral relativism 312
Morgenbesser, Sidney 302
Muggeridge, Malcolm 304
Muller, H. J. 139
Murphy, Nancey 302
Mysteries of Life in Children's Literature, The 327
Mystery of Life's Origin, The 159

N
National Association of Biology Teachers (NABT) 85, 94
National Center for Science Education (NCSE) 139
National Institute of Co-ordinated Experiments (N.I.C.E.) 19, 25, 32, 242, 266, 275
National Science Teachers Association (NSTA) 85, 92
Naturalism 184, 201
  defined 221

Natural law 293
Natural selection
  Balfour's critique of 129
  Bergsonian critique of 124, 134
  challenged by causal principle 189
  challenged by human nature 128
  Chesterton's critique of 128
  denied by human rationality 130, 155, 192, 194, 223
  elastic term 112
  elimination in human societies 28
  leads to moral relativism 132
  replaced by deliberate selection 255
  substitute for intelligent design 137
  unguided process 122, 133, 162, 164, 186
Natural theology 202
Natural Theology 82
Neo-Darwinism 162
New Age spirituality 48
Newman, John Henry 47, 295
Newton, Isaac 228
Nicomachean Ethics 80
Nietzsche, Friedrich 294
Nineteen Eight-Four 272
No Final Conflict 89

O
Ockham's razor 63, 64, 324
Olson, Richard 47
Open Society and Its Enemies, The 140
Origin of Species, The 131, 210
Orwell, George 272
Out of the Silent Planet 79, 242, 245, 323
Oxford University Socratic Club 124, 171

P
Paley, William 82, 123, 159
Parsons, Keith 220
Penelope, Sister 114
Perelandra 120, 121, 134, 242
Peterson, Michael 110, 130, 139, 157, 165
Phenix, Philip H. 302
Phenomenon of Man 135

Philosophia Christi 215, 224
Philosophy
  Lewis's attitude toward 69
Pianka, Eric 37
Pieper, Joseph 318
Pilgrim's Regress, The 25, 31, 48
Piltdown forgery 28, 114
Plantinga, Alvin 183, 190, 224
Plato 269, 295, 310, 316, 320
Polanyi, Michael 196, 299, 304
Polkinghorne, John 133
Pollack, Robert 300
Pope Benedict 316, 335
Popper, Karl 140
Possible Worlds 71
Pratt, J. B. 211
Princess and Curdie 329
Problem of Pain, The 90, 113, 117, 120, 121, 128, 154

Q
Quine, W. V. 229

R
Rationalist Press Association 11
Rea, Michael 225
Reflections on the Psalms 87
"Religion and Science" 55
Reppert, Victor 183, 188, 196
Roberts, Mark 312
Rue, Loyal 23
Russell, Bertrand 251, 303

S
Saving Darwin 116
Saving the Appearances: A Study in Idolatry 64
Savulescu, Julian 235
Schaeffer, Francis A. 89
Science
  abuse of 75
Scientific methodology 63
Scientific reductionism 54
Scientific utopianism 78
Scientism 10, 78, 311
  dangers of 239, 242, 310, 317
  defined 12, 309

destroys discourse 316
impact on humanity 315
teaching of 311
Scientocracy 12
Scopes trial 109
Scott, Eugenie 139
Screwtape Letters, The 249
Sebelius, Kathleen 37
Senior, John 327
Shaw, George Bernard 134, 238, 245
Shermer, Michael 21
Signature in the Cell 158, 163
Silence, role of 334
Silver, Lee 36
Skeptical threat argument 214
Skinner, B. F. 301
Socrates 295
Sparks, Kenton 110
Spencer, Herbert 109, 128
Spirits in Bondage 153, 154
Spong, John Shelby 115
Stapledon, Olaf 238, 243, 246
Steno, Nicholas 72
Stern, Fritz 298
Studies in Medieval and Renaissance Literature 60
Studies in Words 71, 81, 83
Surprised by Joy 25, 125, 154
Swift, Jonathan 301
Synderman, Nancy 36

T
Tao 75, 239, 240, 243, 259, 268
Tattersall, Ian 301
Tending the Heart of Virtue 327
That Hideous Strength 9, 19, 24, 25, 32, 33, 48, 56, 63, 79, 144, 239, 240, 247, 265, 274
Theism 203 223
Theism and Humanism 129, 188, 211
Till We Have Faces 328
Time Machine, The 21
Tirosh-Samuelson, Hava 256
Tolkien, J. R. R. 96, 118, 329

Transhumanism 250
clash with Judeo-Christianity 259
critics of 256
roots of 251
Twins, magic and science as 9, 19, 260

U
Uehiro Center for Practical Ethics 235
Unveiling of the Fall, The 116

V
Virgil 295
Vogt, Karl 47
Voyage of the Dawn Treader 319

W
Waddington, C. H. 243, 283
Wade Center 13
Wallace, Alfred Russel 128
Warrant and Proper Function 224
Watson, David 137
Watson, D. M. S. 92
Wells, H. G. 20, 30, 79, 154, 238, 243, 245
Where the Conflict Really Lies: Science, Religion, & Naturalism 190
Whitehead, Alfred North 48, 49, 77, 298
William of Ockham 64
"Willing Slaves of the Welfare State" 34
"World's Last Night, The" 56

CPSIA information can be obtained at www.ICGtesting.com
Printed in the USA
BVOW031630190712

295700BV00002B/2/P